Routledge Revivals

Geomorphological Hazards in Los Angeles

This book, first published in 1984, deals authoritatively with the nature and management of slope failures and sediment movement and their impact on the hazardous landscape of Los Angeles county. Bringing together for the first time a wide range of information derived from field observations, interviews, manuscript records, local agency reports and published sources, the book presents an analysis of the ways in which a rapidly developing metropolis has come to terms with complex geomorphological hazards. In particular, the events accompanying the major storms of 1914, 1934, 1969 and 1978 are reconstructed in detail.

Geomorphological Hazards in Los Angeles

A Study of Slope and Sediment in a Metropolitan County

R.U. Cooke

Routledge
Taylor & Francis Group
Routledge Revivals

First published in 1984
by George Allen & Unwin (Publishers) Ltd.

This edition first published in 2021 by Routledge
2 Park Square, Milton Park, Abingdon, Oxon, OX14 4RN
and by Routledge
605 Third Avenue, New York, NY 10017

Routledge is an imprint of the Taylor & Francis Group, an informa business

A Library of Congress record exists under LCCN: 84009234

ISBN 13: 978-0-367-76812-6 (hbk)
ISBN 13: 978-1-003-16864-5 (ebk)

DOI: 10.4324/9781003168645

GEOMORPHOLOGICAL HAZARDS IN LOS ANGELES

A study of slope and sediment problems in a metropolitan county

R. U. Cooke
University College London

London
GEORGE ALLEN & UNWIN
Boston Sydney

George Allen & Unwin (Publishers) Ltd.
40 Museum Street, London WC1A 1LU, UK

George Allen & Unwin (Publishers) Ltd,
Park Lane, Hemel Hempstead, Herts HP2 4TE, UK

Allen & Unwin Inc.,
Fifty Cross Street, Winchester, Mass. 01890, USA

George Allen & Unwin Australia Pty Ltd,
8 Napier Street, North Sydney, NSW 2060, Australia

First published in 1984

ISSN 0261–0485

British Library Cataloguing in Publication Data

Cooke, R. U.
Geomorphological hazards in Los Angeles.
—(London research series in geography,
ISSN 0261-0485;7)
1. Landslides—California—Los Angeles
(County)
I. Title II. Series
551.3′03′0979493 QE599.U5
ISBN 0-04-551090-3

Library of Congress Cataloging in Publication Data

Cooke, Ronald U.
Geomorphological hazards in Los Angeles.
(The London research series in geography ; 7)
Bibliography: p.
Includes index.
1. Slopes (Physical geography)—California—Los Angeles County. 2. Landslides—California—Los Angeles County. 3. Earth movements—California—Los Angeles County. I. Title. II. Series.
GB448.C66 1984 363.3′49 84-9234
ISBN 0-04-0551090-3 (alk. paper)

Set in 10 on 12 point Bembo by Preface Ltd, Salisbury, Wilts
and printed in Great Britain by Mackays of Chatham

To
those in Los Angeles County who
manage geomorphological hazards

MOTHER SAVED FROM RUBBLE

2 Children Die as Mudslide Crushes Highland Park Home

BY DOROTHY TOWNSEND

"My babies are in the front room" shouted Mrs. John Gonzalez, pinned

wreckage and five feet of mud that had poured into the house. Fire Department rescuers arrived

Los Angeles Times

FINAL

ONE OF THE WORLD'S GREAT NEWSPAPERS

VOL. LXXXVIII SIXTEEN SECTIONS—SECTION A CC SUNDAY, JANUARY 26, 1969 412 PAGES SUNDAY 35c

MUDSLIDES KILL 9; RAIN DAMAGE GROWS

6,000 Evacuated in Worst Southland Storm Since 1938

BY TED THACKREY JR

Los Angeles Times

VOL. LXXXVIII EIGHT PARTS—PART ONE CC THURSDAY MORNING, FEBRUARY 13, 1969 134 PAGES DAILY 10c

Earthslide Buries Pomona Freeway

Los Angeles Times

VOL. LXXXVIII FIVE PARTS—PART ONE CC WEDNESDAY MORNING, FEBRUARY 26, 1969 94 PAGES DAILY 10c

5 Storm Refugees Known Dead as Mudslide Buries 17

7 HOUSES LOST

Woman Watches in Tears as Home Plunges Into Wash

BY CHARLES HILLINGER

Marilyn Skates stood near the edge of the cliff shivering in the chill of the cold rain—waiting for her home to be washed down rampaging Tujunga Wash early Tuesday morning.

No Sound Heard From Others in Silverado Canyon

BY RICHARD WEST

Mudslides buried alive about 17 storm refugees in Orange County, killing at least five of them, and brought death to a father and three of his children at Mt. Baldy Tuesday.

It was feared that the others buried by the mudslide in the Silverado Canyon area of Orange County were also dead. About 20 persons

Frontispiece Headline images of geomorphological hazards in Los Angeles. (Copyright 1969, *Los Angeles Times*; reprinted by permission.)

Preface and acknowledgements

Metropolitan Los Angeles is one of the great urban phenomena of the 20th century. Its spectacularly rapid growth has been accompanied by a kaleidoscope of changes and challenges that has stimulated research by generations of students, many of them fascinated visitors. This contribution represents my efforts to understand one theme in the remarkable evolution of the metropolis.

Within the turmoil of urban growth, there has been a continuing conflict between the urban community and its highly distinctive physical environment, a conflict that emerges into the spotlight of public concern at times of hazardous physical events such as fires, storms and earthquakes. The purpose of this study is to examine one important aspect of the conflict: the relationships of the geomorphological hazards of slope failure, slope erosion, and sediment transport in floods, and the responses of the metropolis to them. This aspect has been the concern of many local planners, engineers, geologists, politicians and others for decades. It would be presumptuous of me to comment on it without good reasons, and I have at least two.

The first relates to the discovery of data. I have been aware of the long-standing conflict since I first visited Los Angeles in 1964 and mapped the geomorphology of the southern flanks of the Santa Monica Mountains. But my interest in it arises more directly from my work with Richard W. Reeves on *Arroyos and environmental change in the American South-west* (Oxford University Press, 1976) in which we unearthed and analysed a range of archival sources that illuminated the nature of soil erosion and gullying since the mid-19th century in southern California and Arizona. That work gave a glimpse of the archival material relevant to the urban theme of this book. Subsequently, I was fortunate to find a rich harvest of unpublished documents and data, scattered in archives and offices throughout the region, that not only provides a fundamental historical context for studying the urban geomorphological hazards, but also assists hazard prediction and management. This study draws heavily on that harvest. Unpublished sources are supplemented by a prolific crop of published information produced by state, federal and other professional interests, much of which is not readily accessible, especially outside of the United States. The collection of sources I have used is substantial, but it is not comprehensive: for example, there are references in the archives to hazard data and reports that I have been unable to trace. These diverse sources have been augmented by original data derived from field observations, questionnaire surveys and, most importantly, discussions with many of those in the region who have responsibilities for managing geomorphological hazards.

Another reason for this investigation lies in the fact that responsibility for managing the geomorphological hazards has been divided up amongst many different agencies. As a result, no one agency has the task of investigating as a whole the ways in which the conflict has evolved and become absorbed into the fabric of the urban system. Yet the task is useful because it may help busy local environmental managers and communities to gain broader perspectives on their pressing day-to-day problems. Furthermore, the Los Angeles experience in managing geomorphological problems is certainly in advance of that in most democratic urban communities elsewhere, and this study may provide for a wider audience an instructive illustration of how other cities could cope with similar problems. The need to disseminate the fruits of significant hazard experience was emphasised in White and Haas' *Assessment of research on natural hazards* (MIT Press, 1975). They advocated detailed studies that, *inter alia*, examine what happens when disasters occur, explore the development of hazard responses in specific communities over long periods, and facilitate the exchange of information amongst agencies. This study is directed towards these objectives.

A prodigious quantity of relevant data has forced me to restrict the scope of the enquiry. In terms of hazards, I have focused on slope failure, slope erosion, and sediment movement as comprising a distinctive group of related fluvial processes that link with my hazard studies in other relatively dry environments. The broader problem of flooding has already received much attention, notably in R. Bigger's *Flood control in metropolitan Los Angeles* (UC Press, 1959); and other geomorphological problems in the Los Angeles region, arising, for example, from coastal processes and oil extraction, differ fundamentally from the fluvial processes described here. Spatially, the research is confined to Los Angeles County – the smallest unit large enough to include the full range of fluvial hazards and management agencies. Temporally, the study extends from the catalytic storms of 1914 to the stormy winter of 1978 – a period long enough to allow recognition of secular trends and to include the principal geomorphological crises of the 20th century. However, 1978 is not a natural ending: 1980 brought further storms, and as I write the latest storms are destroying Santa Monica pier.

Many professionals directly involved with the hazards and their management have generously assisted my research. During the time I spent with them I came to appreciate both the difficulties of their jobs and the substantial success they have achieved, often against enormous odds. Those who have helped me are too numerous to name individually, but they include employees of most of the management agencies mentioned in the text, together with a number of academics, consultants, politicians, journalists and private citizens. I have imposed particularly heavily on the staff of the Los Angeles County Flood Control District and several agencies of the City of Los

Angeles. I do not suppose my essay will receive universal approval locally, for it inevitably touches on controversial and sensitive matters and it may suffer from sins of omission, misplaced emphasis, and misinterpretation which experienced local managers and consultants will recognise at once. Thus, while I am greatly indebted to those who have assisted me, for the study would have been impossible without their friendly cooperation, I am alone responsible for the views expressed.

My enquiry was initially funded (in 1973) by the American Council of Learned Societies. The Department of Geography at UCLA kindly provided a *pied-à-terre*, and Howard Nelson gave munificent encouragement. Robin Flowerdew assisted in the analysis of data for the 1969 storms. Most of the diagrams were drawn by Kenneth Wass in the Department of Geography at University College London. Barbara Cooke devotedly typed the manuscript and packed the lunches. Edmund Penning-Rowsell and Joseph Cobarrubias kindly provided helpful comments on the penultimate draft. I am grateful to them all.

I am pleased to acknowledge the following for permission to reproduce copyright material: The *Los Angeles Times* for the frontispiece and Fig. 2.12; the Regents of the University of California for Fig. 1.9A; H. J. Nelson for Figs 1.9B and 1.17A and F; the Association of American Geographers for Fig. 1.18; and J. Cobarrubias for Fig. 2.11.

R. U. COOKE
London
January 1983

Contents

Abbreviations used in the text

CBL Los Angeles County Building Laws
CLABC City of Los Angeles Building Code
LACFCD Los Angeles County Flood Control District
LACDA Los Angeles County Drainage Area Project
LARFP Los Angeles River Flood Prevention Project
UBC Uniform Building Code
USACE US Army Corps of Engineers
USDA US Department of Agriculture
USDAFS US Department of Agriculture Forest Service
USGS US Geological Survey (Department of the Interior)
USDHUD US Department of Housing and Urban Development

1

CONTEXTS

Geomorphological hazards

The major geomorphological hazards in Los Angeles County (and indeed throughout southern California) are associated with valley-side slopes and channels in the mountains, and with the drainage channels and alluvial plains beyond the mountains. The principal components of these natural geomorphological systems are shown schematically in Figure 1.1. The valley-side slope activity includes three main processes: rock decomposition, and the movement of material downslope by slope erosion or slope failure. Slope erosion comprises surface movement of debris either under the predominant influence of gravity (dry erosion) or in conjunction with flowing

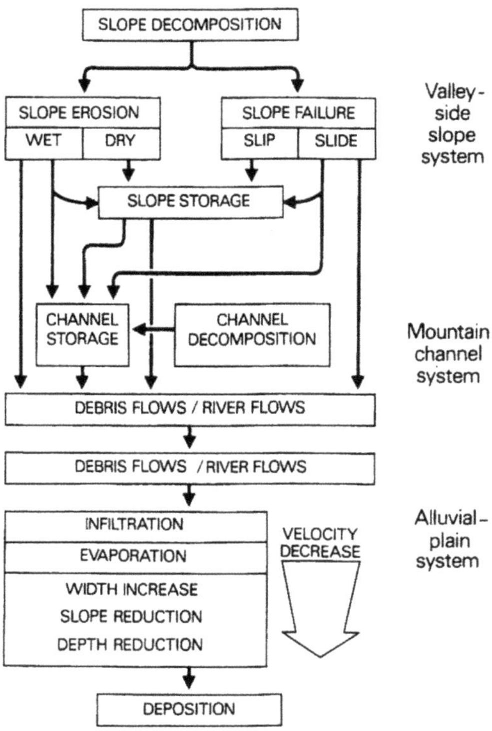

Figure 1.1 Principal features of the fluvial geomorphological systems in Los Angeles County.

DOI: 10.4324/9781003168645-1

water (wet erosion). Slope failure may take many forms, but the most significant are surface-soil slipping and landslides. Both slope erosion and slope failure pose problems for people and property in hilly and mountainous areas. Both are also the main contributors of sediment to the channel network – the second major locus of geomorphological hazards. Hazards associated with the mountain channels and the channelled alluvial plains beyond the mountains arise from the way in which water and sediment are moved through drainage systems in semi-arid areas. The flows are ephemeral, range from river flows low in debris to relatively viscous debris flows, and, unless controlled, often follow relatively unpredictable courses across alluvial slopes. When and where flows leave their channels, the resulting floods of sediment and water have posed problems for plains dwellers in much of Los Angeles County.

The geomorphological processes are natural phenomena that have only become serious hazards because they have increasingly imposed themselves upon a vulnerable, often unsuspecting, and rapidly growing urban community. Part 1 of this book provides a review of three physical contexts of the hazards: the environmental conditions that predetermine geomorphological activity, the dynamics of environmental change that modify the impact of geomorphological processes over time, and the salient features of the geomorphological phenomena themselves. The review reveals that although much is known, the problems facing environmental managers are exceptionally complex, and prediction still involves uncertainty. This conclusion serves as a prelude to an examination (in Part 2) of the crises produced within Los Angeles County by geomorphological processes during the major storm events of the 20th century. There follows in Part 3 a survey of the ways in which the evolving metropolis has adjusted to the geomorphological hazards.

The hazard context

Predetermining conditions

The stage upon which the processes play is largely predetermined by topography and the pattern of drainage, the underlying geology and soils, the vegetation cover, and the external climatic forces (Fig. 1.2). These predetermining conditions are reviewed in this section. But fundamental to an understanding of geomorphological activity in the Los Angeles region is the fact that some of the predetermining conditions and the geomorphological processes themselves are not constant: they evolve as a result of changes in, for instance, the location, frequency, magnitude and duration of storms, tectonic activity and fire, the transformation of surface characteristics by the processes of urbanisation, and hazard management (Fig. 1.2). These changes are

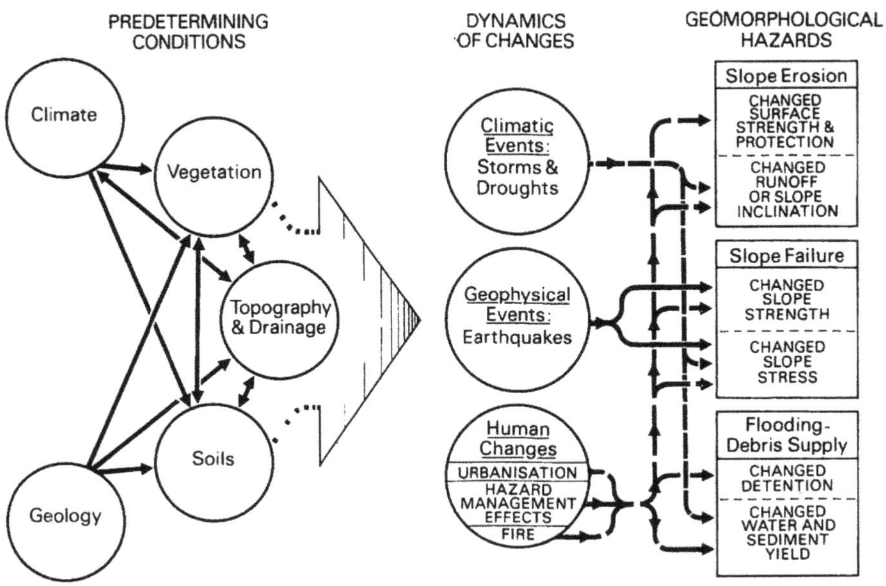

Figure 1.2 Predetermining conditions and dynamics of changes relating to geomorphological hazards in Los Angeles County.

DOI: 10.4324/9781003168645-2

considered in the following section. To avoid splitting an important theme between the two sections, climate is considered as a whole in the next section.

TOPOGRAPHY AND DRAINAGE

Los Angeles County comprises portions of three of California's geomorphological provinces (Miller 1957): the Mojave Desert, the Transverse Ranges (represented by the Santa Monica, the Santa Susana, and the San Gabriel mountains), and the Los Angeles Coastal Plain (Fig. 1.3). Each of these areas incorporates substantial topographical variety.

The Mojave Desert consists (within the county) of alluvial plains extending north from the flanks of the San Gabriel Mountains, mainly in the Antelope Valley. The plains, punctuated by isolated, rocky hills, are periodically flooded by flows from the mountains.

The mountains of the Transverse Ranges are characterised by high relative relief (often over 400 m in small drainage basins), deep and pervasive dissection, and innumerable very steep, rectilinear or convexo-concave slopes. As a result, the valley-side slopes in the mountains are exceptionally active environments in which rates of debris production and removal are extremely rapid by comparison with those of surrounding areas and of different climatic regions.

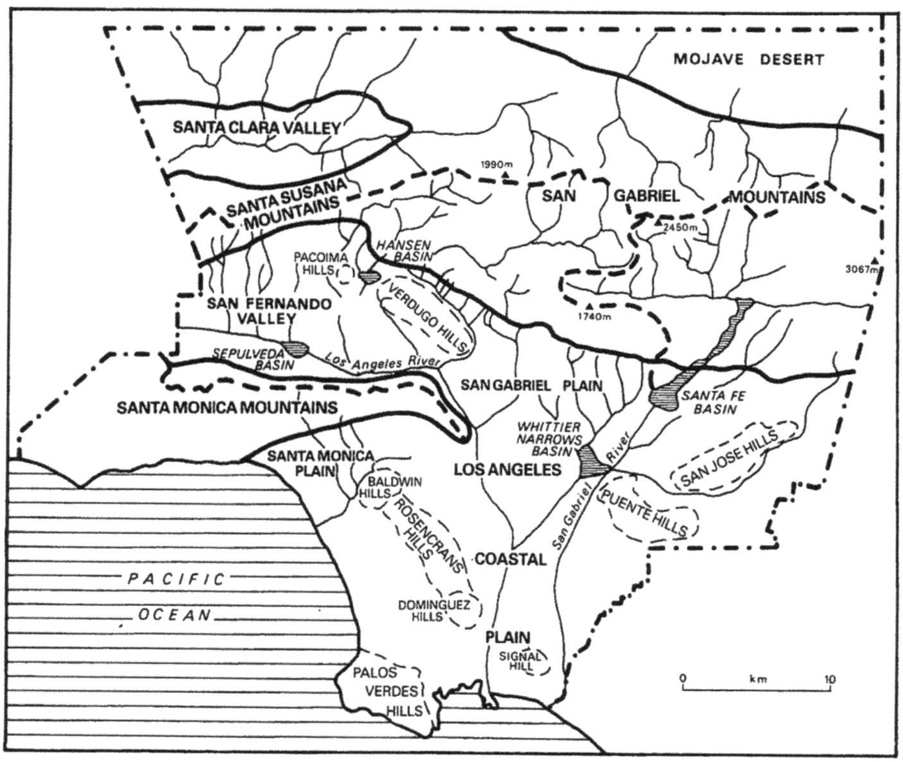

Figure 1.3 The major geomorphological regions of Los Angeles County.

Data on mountain-slope gradients are abundant but not comprehensive. Retzer *et al.* (1951), in an early study of 46 mountain watersheds in the Los Angeles River catchment of the San Gabriel Mountains, gave the *average* slope as 68%, and in 62.2% of the area mapped they found slopes *exceeded* 70%. The US Department of Agriculture Forest Service (USDAFS 1974) showed that in the same catchment, 73% of slopes exceeded 60%, 20% were between 30% and 60%, and only 7% were below 30%. Within the 6885 ha of the San Dimas Experimental Forest in the San Gabriels, Bentley (1961) found that nearly all slopes exceeded 55%, and over a third exceeded 70%. Strahler (1950), in a detailed study of the *steepest* segments of slope profiles in the Verdugo and San Rafael hills, recorded average slope angles in three sample areas of 97%, 90% and 90.1% (with standard deviations of 5.7%, 4.3% and 4.8% respectively). These data emphasise the steepness of mountain slopes and their potential instability, given that many exceed the normal angle of repose of unconsolidated material (c. 70%).

It is characteristic for the mountains to terminate abruptly against their fringing alluvial plains along steep mountain fronts. As a result, drainage from each of most entrenched mountain catchments emerges on to the plains through a channel that feeds an alluvial fan.

The alluvial areas of the Los Angeles Coastal Plain, together with the largely enclosed alluvial basin of San Fernando Valley, are dominated by the marginal alluvial fans, and the valley-floor floodplains of the major rivers. In addition, the plains include several modest upland areas (such as the Puente, Palos Verdes, Dominguez and Baldwin hills) that are islands of bedrock, often raised by relatively recent tectonic movements, in a sea of alluvium and urban sprawl.

Drainage in the region is dominated by the Los Angeles, San Gabriel, and Santa Clara river systems. At present, the heavily engineered Los Angeles River passes along the southern side of the San Fernando Valley, between the Santa Monica Mountains and the Verdugo Hills, south-east across the Los Angeles Plain to Vernon, and then south to San Pedro and the Pacific Ocean (Fig. 1.3). It receives several major tributaries from the north and east – the Pacoima and Tujunga washes in San Fernando Valley, the Verdugo Wash drainage at Glendale, and the Rio Hondo at Southgate. The San Gabriel River rises in the San Gabriel Mountains and flows south into the San Gabriel Valley at the Santa Fé flood-control basin, where the flow is split between the Rio Hondo and the main channel that passes south to Alamitos Bay. While these major rivers act as the ultimate foci of most drainage, it would be wrong to assume that the drainage systems as a whole behave in the same integrated way as major river systems in temperate areas. In reality, the systems are only rarely fully operational and integrated. Drainage is mostly ephemeral, responding to occasional local rainfalls, and the distinctive natural feature of mountain flows coursing down canyons and being disseminated and dissipated on alluvial fans, means that only major storms have generated sufficient runoff to cause floods along the Los Angeles and San Gabriel rivers themselves.

Within the watersheds of the major rivers in the Los Angeles area, there are many small, relatively independent catchments, each of which includes a mountain hinterland and a segment of the alluvial plain, and poses or has posed local problems for urban development. These hazardous catchments (e.g. Fig. 1.4) are found predominantly around the San Fernando Valley and along the margins of the San Gabriel Mountains.

Therefore, one of the keys to understanding the complexity of the geomorphological problems of Los Angeles County lies in the importance of a huge number of small, ephemerally active systems – the innumerable, relatively steep slopes in the mountains and hills, and the numerous, relatively independent drainage basins straddling the mountain–plain boundary. It is these numerous systems that combine to produce difficult regional slope and channel hazards, and collectively contribute water and sediment to the large-scale flooding problems associated with the Los Angeles, San Gabriel and Santa Clara rivers.

GEOLOGY AND SOILS

The geology of Los Angeles County has been studied for many years. Numerous regional and site maps are available. Some are published, but most are in manuscript form, and many are buried in theses and consultancy reports. This huge volume of geological material reflects both the importance of geology to urban development and the intrinsic geological interest of the area.

From the perspective of geomorphological hazards, the most important aspects of geology are lithology (particularly as this relates to weathering rates, erosion, and soil development); the nature of structures and bedding planes (chiefly as they influence slope stability); and the pervasive threat of seismic

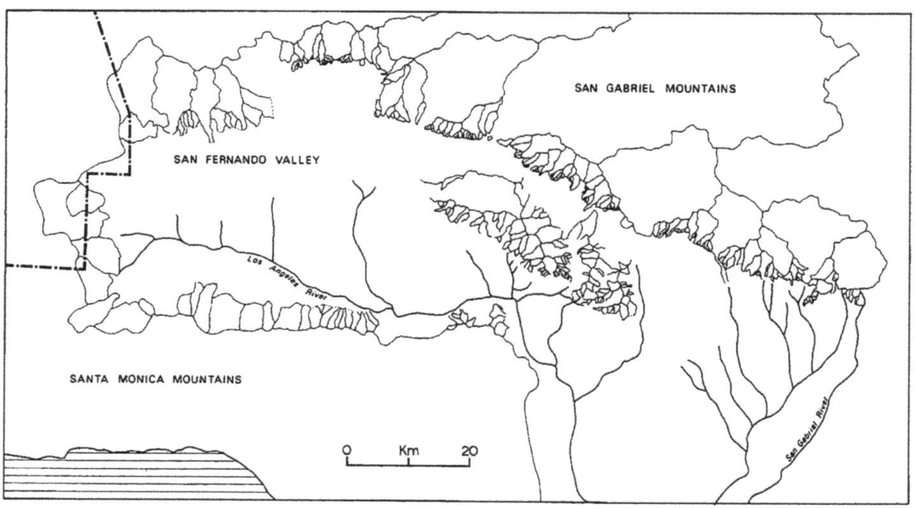

Figure 1.4 Mountain catchments in the Los Angeles River Watershed (after USDAFS 1977).

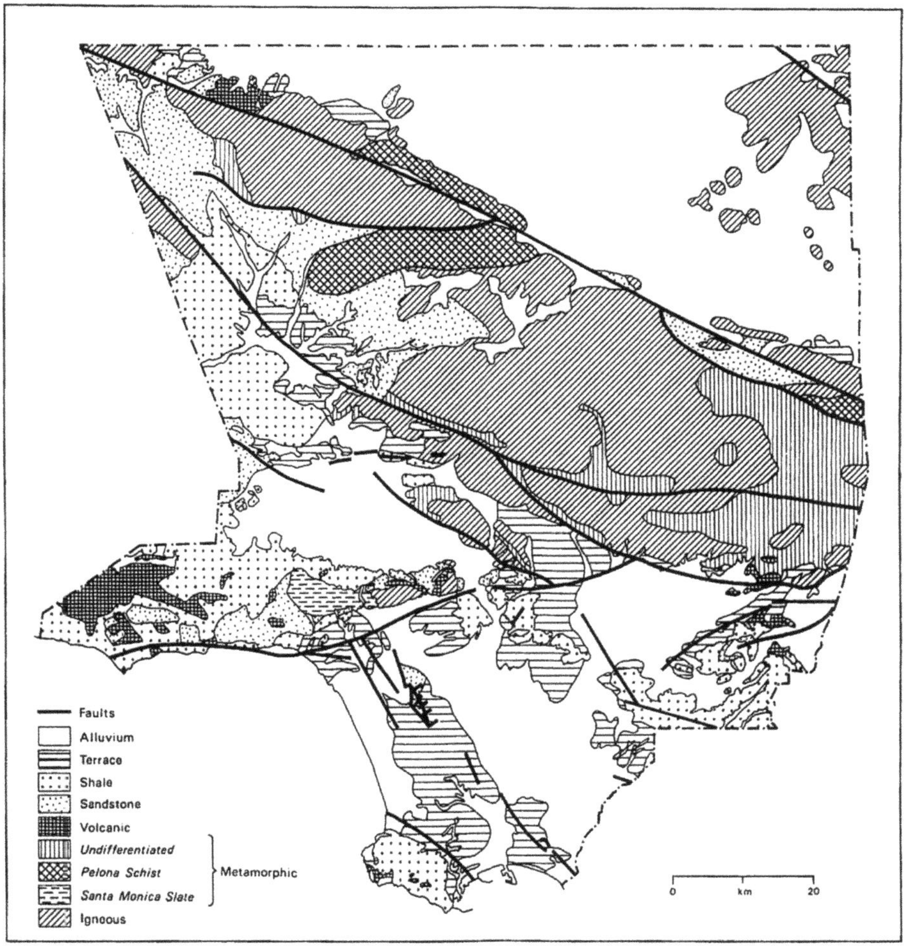

Figure 1.5 Generalised geological map of Los Angeles County (after Los Angeles County, Department of Regional Planning 1974).

activity. Of fundamental importance is the fact that the hills and mountains of Los Angeles County are relatively young, tectonically active, and composed in general of weathered and fractured sedimentary, igneous and metamorphic rocks.

Lithology Figure 1.5 shows a generalised geological map of Los Angeles County (Los Angeles County, Department of Regional Planning 1974) which simplifies the data on other maps (e.g. Jahns 1954, California Division of Mines and Geology, 1969) into a few major lithological groups.

Alluvium and *terrace deposits* comprise Recent and Quaternary sediments that are poorly consolidated sands, gravels, clays and silts derived mainly from mountain floods and marine deposition. From the geomorphological point of view, they are often unstable in steep natural and artificial slopes, relatively permeable (although they frequently include relatively impermeable clay partings), and easily eroded. They are particularly vulnerable to attack where they have been elevated by tectonic movements, as in some areas of the Los Angeles Coastal Plain.

Shale and sandstone represent the two major groups of sedimentary units (mainly of Tertiary age) which dominate the Santa Monica and Santa Susana mountains, the Palos Verdes Hills, and the Puente-San José Hills. The shale group includes the Puente, Pico, Repetto, Modelo and several other Tertiary formations of shales and siltstones which, while better bedded and better consolidated than the Quaternary sediments, are nevertheless also prone to weathering, slope failure and erosion. The sandstone group – found mainly in the Santa Monicas and the Santa Susanas – includes coarser-grained sandstones and conglomerates in the Sespe, Martinez, Chico and other formations that are, because of their greater consolidation and innate strength, broadly more resistant to slope failure and erosion than the shales and the Quaternary sediments.

Volcanic rocks, of Miocene age, including extrusive flows, tuffs, breccias and some intrusions, occur in the mountains, most notably in the higher ground of the western Santa Monicas.

Igneous rocks, principally granite, and anorthosite and other gabbroic rocks, dominate the central areas of the San Gabriels. These coarsely crystalline and heavily fractured rocks disintegrate in a granular fashion, supplying much sand and silt to the major drainage catchments.

Metamorphic rocks, associated with the fringes of the igneous intrusions, dominate the eastern San Gabriel Mountains and the south-central Santa Monica Mountains; the Pelona schist occurs in the north of the county, and the Santa Monica slate (a phyllite) outcrops in the eastern Santa Monicas. In general, the volcanic and igneous rocks appear to be less susceptible to slope failure than the alluvium, shale and sandstone sequences.

Soils, hydrophobicity and weathering General comments by several authors, including Rice (1973), and Bailey and Rice (1969), and the more detailed observations of Retzer *et al.* (1951) in 46 watersheds of the Los Angeles River watershed, all confirm the view that soils are closely related to rock type in the hills and mountains of Los Angeles County. In general, soils in alluvium and the finer-grained sedimentary rocks give the finest-textured soils, whereas those on intrusive igneous rocks tend to be coarse and sandy. Because slopes are often steep and unstable, soils are usually shallow, poorly developed, relatively young and highly permeable. Some, especially on the igneous rocks, are formed on residual weathering mantles that may be metres deep. Such mantles differ little in their hydrological properties from the soils themselves, both often having very high infiltration capacities. For example, studies at the San Dimas Experimental Forest (in the San Gabriel Mountains) showed that of the 300 largest storms occurring at the station in 24 years, in none was the average hydraulic conductivity of the soil exceeded for an hour, and only 2.5% of the precipitation fell at a rate higher than the infiltration capacity (Rice 1973).

Despite this evidence of relatively high infiltration capacities, they are spatially variable and can change over time. Particularly important in this context are water-repellent soils which are widespread in areas of chaparral vegetation. For example, Krammes and DeBano (1965) suggested that 60% of soils in the San Dimas Experimental Forest are water repellent. Water repellency arises primarily through the coating of soil particles with hydrophobic organic substances. It is a phenomenon that is caused, or significantly intensified, by chaparral fires in southern California. Before a fire, organic matter accumulates on the surface, and the upper soil layers become water repellent due to the intermixing of partially decomposed organic matter and mineral soil (Foggin & DeBano 1971, DeBano 1981). During a fire, the surface litter is burned, organic particles coat, and become chemically bonded to, the cooler soil particles below, and vaporized substances condense on mineral soil particles at depth. The new, deeper, thicker and usually discontinuous water-repellent layer alters infiltration capacities: the surface layer is first wetted, but further infiltration is initially extremely low, and rises only slowly during a storm. As a result, surface and near-surface runoff on a burned watershed during a storm are increased, and this leads to accelerated soil erosion (Krammes & Osborn 1969) – a fact of major importance in explaining erosion rates during storms.

The relative rates at which the varied rock types disintegrate under present conditions, creating weathering mantles and soils and nourishing sediment supply, is a matter of fundamental importance that has received scant attention. Studies by the USDAFS (1953), and Scott and Williams (1978) suggested that differences do exist, at least as they are reflected in sediment yields. Scott and Williams (1978) estimated erodibility of soils in the mountains and foothills of the Transverse Ranges using the dispersion ratio and the surface–aggregation ratio. Table 1.1 shows a sample of their results for the major lithological types that occur in Los Angeles County. The study revealed differences that are probably reflected in variations in the rate of debris supply to drainage systems.

Structures, joints and bedding planes Figure 1.5 shows the major faults that transect the area. These faults, which include those of the San Andreas and Santa Susana systems, have been studied in detail (e.g. Jahns 1954); their geomorphological importance lies in the facts that many are active (see the discussion on earthquakes, p. 15), they delimit the main mountain areas, they often provide relatively easily eroded avenues within the mountains, and they are accompanied by myriads of minor faults and joints that are vulnerable to erosion and may facilitate slope failure.

Table 1.1 Dispersion and surface–aggregation ratios for various soils on the main geological units in the mountains and foothills.

		Dispersion ratio – mean confidence limit at 0.95% probability	Surface–aggregation ratio – mean confidence limit at 0.95% probability
I	Quaternary alluvium	56 ± 18	132 ± 76
II	shale and sandstone groups e.g.		
	(i) Pliocene and non-marine sediments	53 ± 4	121 ± 12
	(ii) Upper Miocene non-marine sediments	44 ± 3	69 ± 9
III	volcanic rocks	52 ± 6	140 ± 42
IV	igneous rocks (granite)	36 ± 6	121 ± 35
V	metamorphic rocks (pre-Cretaceous marine meta-sediments)	47 ± 5	139 ± 30

Note: the dispersion ratio (DR), expressed as a percentage, is the ratio of the percentage of measured silt-sized and clay-sized particles in an undispersed soil to the percentage of the same sizes after dispersion; the surface–aggregation ratio (SAR), expressed as a percentage, is the ratio of the surface area of particles coarser than silt divided by the value of the aggregated silt plus clay.

Source: data selected from Scott and Williams (1978).

VEGETATION

Vegetation types and patterns The 'natural' vegetation of Los Angeles County comprises five principal types: desert scrub, sagebrush, grassland, woodland and forest and, most important areally and geomorphologically, chaparral. The pattern of vegetation made by these types (Fig. 1.6) is broadly one of altitudinally arranged zones reflecting local climates (see Fig. 1.9, p. 18).

Desert scrub, which occurs on the alluvial plains of the Mojave Desert and the lower flanks of the San Gabriels, is a community of scattered trees and shrubs. The most common tree is the Joshua tree (*Yucca brevifolia*), and common shrubs include creosote brush (*Larrea tridentata*) and burroweed (*Franseria dumosa*). From the geomorphological viewpoint, this sparse and stunted vegetation leaves a relatively high proportion of the ground surface exposed directly to the action of water and wind, and contributes towards relatively poor binding properties and low infiltration capacities for the soil (e.g. Cleveland 1971a).

Sagebrush communities occur mainly on the coastal flanks of the Santa Monica Mountains and Palos Verdes Hills, and on the lower slopes surrounding San Fernando Valley, generally up to about 500 m above sea level (asl). The stands of sagebrush (*Artemesia californica*) and subsidiary genera of sage (*Salvia*), California buckwheat (*Eriogonum*), *Haplopappus* and *Encelia* (Hanes 1971) form relatively open stands, with substantial proportions of bare ground in summer and more complete cover in winter. There is a tendency for grassland to be established, especially after fire.

Grassland (most of which is the creation of man) is found amongst the

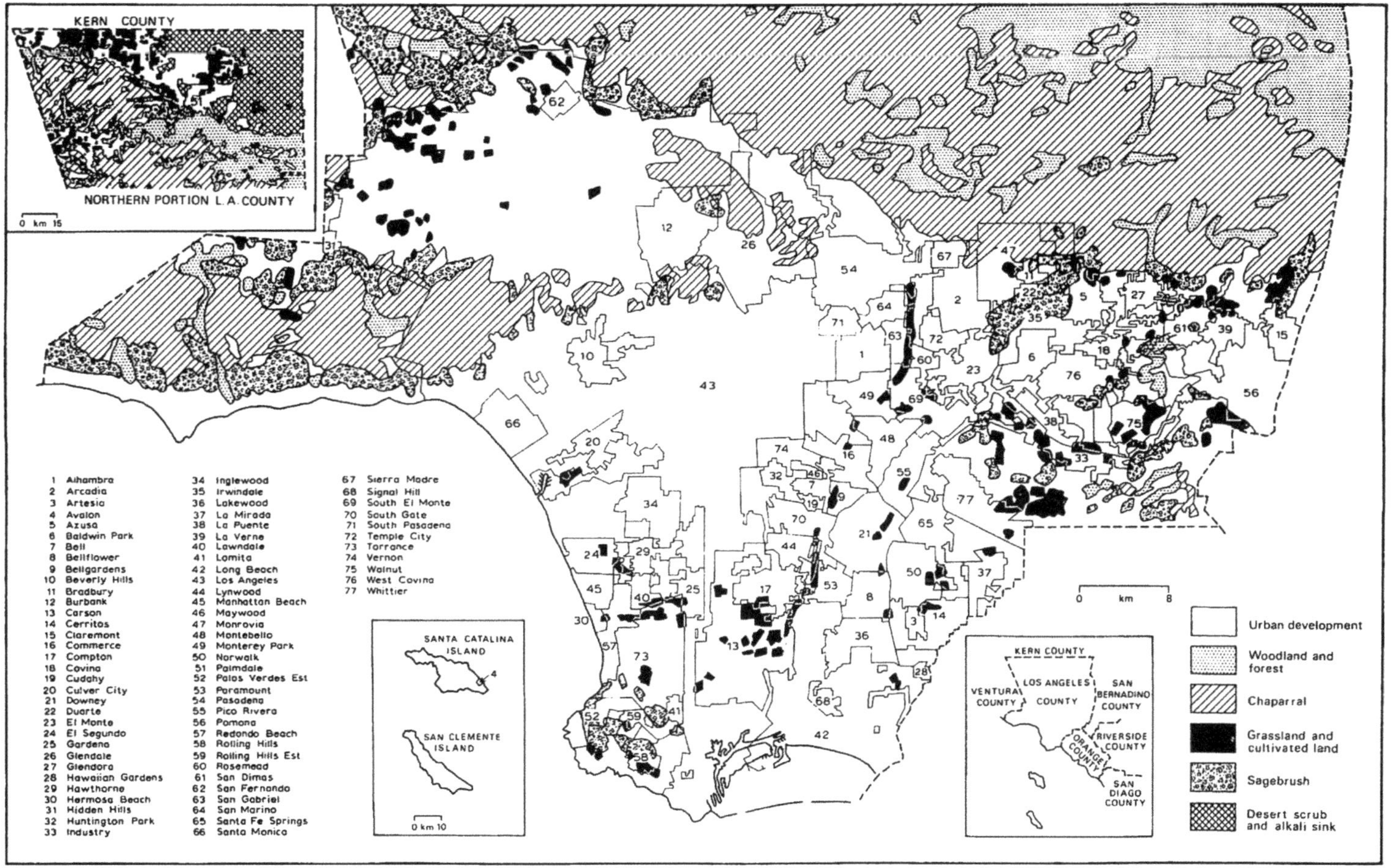

Figure 1.6 Major vegetation zones in Los Angeles County (after Los Angeles County, Regional Planning Commission 1973).

sagebrush areas and, more extensively, in those lowland and hilly areas within them that are beyond the margins of urban and agricultural development. Grass growth is sometimes stimulated after fires by artificial seeding, in the belief that grass cover provides a better protection against erosion (e.g. Rowe & Reimann 1961). Grassland communities have higher infiltration capacities than some other vegetation types, and the strength of their root systems in binding soil is low, so that soils can become saturated relatively easily, and superficial slope failure is common (Cleveland 1971a).

Chaparral – the most extensive vegetation type – is found in the mountains and hills above the sagebrush communities. The shrubs and trees are 1–6 m tall, and characteristically have evergreen, hard, relatively small leaves. The principal species include chamise (*Adenostoma fasciculatum*) and, quantitatively less common, oaks (*Quercus dumosa agrifolia*), *Arctostaphylos, Ceanothus, Heteromeles, Rhamnus, Cercocarpus, Prunus, Yucca* and laurel (*Rhus laurina*) – (Hanes 1971). The chaparral cover is very much determined by fire history (see p. 22), but at its climax it can be extremely dense and impenetrable, offering protection to the partially bare surface from both sunlight and rainfall (Bailey 1954).

Surface protection offered by chaparral vegetation is not uniform (Retzer *et al.* 1951). For example, chamise provides poor protection (producing little litter and low cover), whereas oaks, especially in the early stages of their development, effectively shield a surface that is relatively free of ground cover and may be susceptible to creep and wash processes. In contrast, buckthorn (*Ceanothus*) provides excellent surface protection (Retzer *et al.* 1951), with a dense canopy and a thick litter.

Woodland and forest only occur extensively at the highest altitudes in the San Gabriel Mountains. Here (above about 1500 m) are found pine forests, dominated by *Pinus*, together with *Libocedrus, Pseudotsuga, Quercus* and other genera. On the northern desert flanks of the range the pine forests yield to *pinyon–juniper woodland*, an open, dwarf coniferous forest, comprising pine-nut trees (*Pinus monophylla*) and juniper (*Juniperus californica*), and this in turn merges into the Joshua tree and desert shrub communities at lower elevations in the desert. The pinyon–juniper woodland offers poorer ground protection than the pine forests, and can be converted into chaparral by fire, which eliminates the dominant woodland species (Hanes 1971). Woodland communities are also found on lower-level tongues extending down some canyons to the south of the San Gabriel watershed.

Within these broadly altitudinal zones, local variations in the density and nature of vegetation communities are determined largely by fire history (see p. 27), aspect and soil conditions. The best locations, suffering least from summer drought, are those of north-facing slopes with thicker soils (e.g. Retzer *et al.* 1951). By implication, such sites are likely to be less vulnerable to surface erosion and to shallow slope failures. The altitudinal zonation of vegetation is further complicated by riparian vegetation along canyon floors, where greater availability of water often promotes the growth of trees and shrubs which, in turn, act as obstacles to debris movement.

Vegetation and erosion Vegetation cover is a fundamental factor predetermining the resistance of the ground surface to geomorphological processes. Its principal rôles are in protecting and binding the surface, and influencing both water and sediment yields. The different vegetation communities in the Los Angeles region play these rôles in different ways, and with different effects. While their general contribution is understood, detailed studies have largely been confined to grassland and chaparral communities – most notably in the San Dimas Experimental Forest.

The effect of vegetation on rainfall – the process of interception, and redistribution by throughfall, stemflow or evaporation – is critical to understanding geomorphological processes. Interception loss reduces the availability of water for erosion and runoff; stemflow may, in some circumstances, concentrate runoff and provide loci for erosion; and throughfall (including leaf drip) may modify raindrop impact and thus impact erosion. The rôle of surface litter is also important in this context for it can store water, reduce soil-water evaporation, reduce raindrop impact, retard runoff, and promote water repellency. Studies of these phenomena at the San Dimas Experimental Forest (e.g. Hamilton & Rowe 1949, Patric 1959, Corbett & Crouse 1968) have described and compared the rôles of chaparral and grassland in influencing water yield, increasing forage, and improving fire control. For example, Corbett and Crouse (1968) showed that chaparral has a higher annual net rainfall interception than annual grass, mainly because the former functions as an efficient interceptor throughout the year, whereas grass interception is high only in the later stages of growth at the end of the rainy season. Furthermore, interception losses for grass and chaparral were proportionately higher for small storms when warming temperatures stimulate growth, than for large storms (Table 1.2). Rowe and Reimann (1961), qualifying these ideas, demonstrated that the conversion from chaparral to grass would only increase water yield in wet years on deeper soils.

The geomorphological consequences of such differences have not been fully explored. The management aim of increasing water yield by vegetation conversion may not be compatible with a second aim of reducing erosion and

Table 1.2 Interception losses at San Dimas Experimental Forest (Corbett & Crouse 1968).

	Hydrologic year	Interception loss (%)		Total (mm)
		grass	chaparral	
(a) Small storms, low rainfall	1963–4	13.8	16.6	381–812
(b) Large storms, high rainfall	1953–4	5.0	10.9	609–2336

flooding. Rowe and Reimann (1961) indicated that the chaparral–grass transformation, once completed, should reduce soil loss by erosion (presumably by reducing raindrop impact, restricting runoff on soil surfaces, and improving soil resistance to erosion). But this laudable objective may conflict with the possible effects of vegetation removal and transformation on slope failure. Bailey and Rice (1969), and Corbett and Rice (1966) indicated that the occurrence of soils slips at experimental sites in the San Dimas Experimental Forest was closely related to the rooting habit and density of vegetation (Table 1.3). For example, shallow, laterally spreading roots characterise sagebrush and grass, and, in general, surfaces with these vegetation types were much more susceptible to failure than those covered by chaparral. In addition, these studies showed that the higher density of shallow-rooted annual grasses apparently improves slope stability by comparison with lower-density, deeper-rooted perennial grasses in the context of surface failure, and slope failure diminishes as the height and the density of chaparral increases. Vegetation that includes trees with deep tap roots (e.g. *Quercus dumosa*) is associated with surfaces less susceptible to soil slips. What is true of soil slips may also be true of soil creep, although it is probably less true of deep-seated slope failures that occur below maximum root depths (Retzer *et al.* 1951).

The consequences of vegetation conversion may also extend into the channel system for, as Orme and Bailey (1970, 1971) showed, changed discharge of water and sediment arising from replacement of chaparral by grass alters channel patterns of erosion and deposition, and may lead to channel widening, deepening, and reduction in gradient.

Dynamics of change

The geomorphological processes at work on the hills, mountains and plains of Los Angeles County are not constant and they do not attack an unchanging landscape. The processes depend mainly on natural forces arising from geophysical and climatic events. And it is fundamental for an understanding of the geomorphological hazards and the problems of their management in Los Angeles to appreciate that the landscape has been profoundly and dramatically transformed in the last 100 years or so as a result of human activity. Two groups of transformations are particularly important: the transformation of natural vegetation and agricultural land into the fabric of urban communities, and the periodic destruction of vegetation by fire. Both of these changes (the former normally permanent, the latter temporary) radically alter the response of the landscape over time to geophysical and climatic events. A further group of important changes arises from deliberate modifications to the geomorphological system resulting from hazard management decisions.

This section briefly reviews the natural forces and the nature of secular changes in the landscape.

Table 1.3 Percentage of area of vegetation types that slipped on the Bell Canyon watersheds, San Dimas Experimental Forest, California, 1966–7 (Bailey & Rice 1969).

Vegetation type	Percentage of area in slips
sage and barren	23.9
perennial grass	11.9
annual grass	6.5
riparian woodland	5.3
chamise chaparral	3.3
oak chaparral	2.6
broadleaf chaparral	1.2

EARTHQUAKES AND TECTONIC MOVEMENTS

In the context of the long-term landform evolution of the Los Angeles region, tectonic activity plays a fundamental rôle in altering the altitudinal relationships between land and sea. There is geomorphological evidence (e.g. Putnam 1942, Sharp 1954) that the mountains were raised spasmodically throughout the Quaternary, often at a relatively rapid rate, and such changes are ultimately responsible for their steep slopes, deep dissection, and high relative relief. Precise data on rates of uplift are few (Table 1.4), but they show rates that are consistently higher than predicted gross rates of denudation. Scott and Williams (1978) suggested that high denudation rates in the Transverse Ranges, for example, may be about 2.3 m per 1000 years (per 1.3 km^2 drainage area). Such high rates, like those reported by Schumm (1963), are generally less than the estimated rates of tectonic uplift. Where rates of uplift exceed rates of denudation – as appears to be the case in the hills and mountains of Los Angeles County – gravitational force acting through steep slopes enhances mass movement of sediments and encourages relatively high debris yields.

Earthquakes (the local and most tangible expression of tectonic activity) are serious hazards in their own right in the Los Angeles region (Richter 1954),

Table 1.4 Estimated rates of tectonic uplift in southern California.

Location	Rate (m/1000 years)	Evidence	Source
Baldwin Hills	(a) 5–8	stratigraphic dating	Bandy & Marincovich (1973)
	(b) 9.12	levelling surveys	Gilluly (1949)
Transverse Ranges	21	fault dislocation of Holocene rocks	Scott & Williams (1978)
Santa Monica Mtns (flanks)	4	levelling surveys	Stone (1961)
San Gabriel Mtns (southern flanks)	6	levelling surveys	Stone (1961)
Cajon Station	4.34	levelling surveys	Gilluly (1949)

where folk memories of the Long Beach disaster in 1933 and the more recent San Fernando Valley earthquake are still clear (e.g. *California Geology* 1971, Los Angeles County Board of Supervisors 1971). Geomorphologically, earthquakes can produce ground settling and they can trigger slope failures. Evidence of such failures is not normally recorded systematically, but Morton's (1971) examination of aerial photographs and field damage showed that many slope failures occurred in a 250-km^2 area of the mountains in the vicinity of the San Fernando Valley earthquake on 9 February 1971 (Fig. 1.7). These failures included fairly superficial features such as rock falls, soil falls, debris slides, avalanches and slumps, but not, perhaps significantly, more deep-seated and complex landslides.

Earthquake hazards have troubled Los Angeles for many years (e.g. Los Angeles Chamber of Commerce 1933), but the enhanced perception of the problems following the San Fernando disaster led to a flurry of research into predicting the location of seismically sensitive areas, including areas susceptible to earthquake-triggered landslides (e.g. Radbruch & Crowther 1973). This work led, *inter alia*, to the preparation of a seismic zoning map of Los Angeles County (Los Angeles County, Department of Regional Planning 1974, various reports), which is based on the interpretation of geological, landslide, shallow-

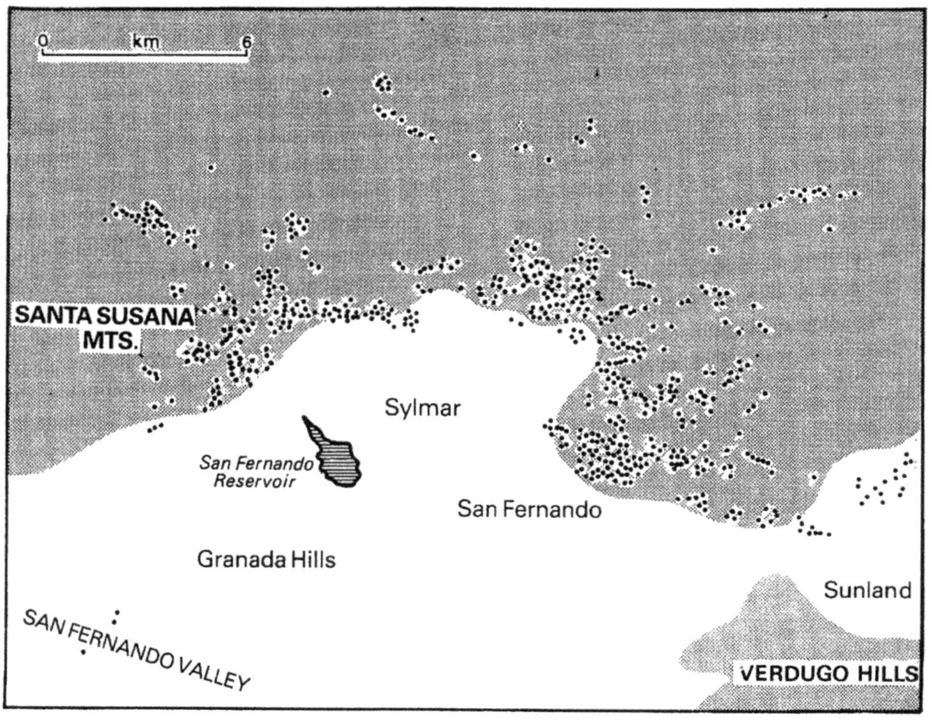

Figure 1.7 Mountain-slope failures associated with the San Fernando Valley earthquake, 9 February 1971 (after Morton 1971).

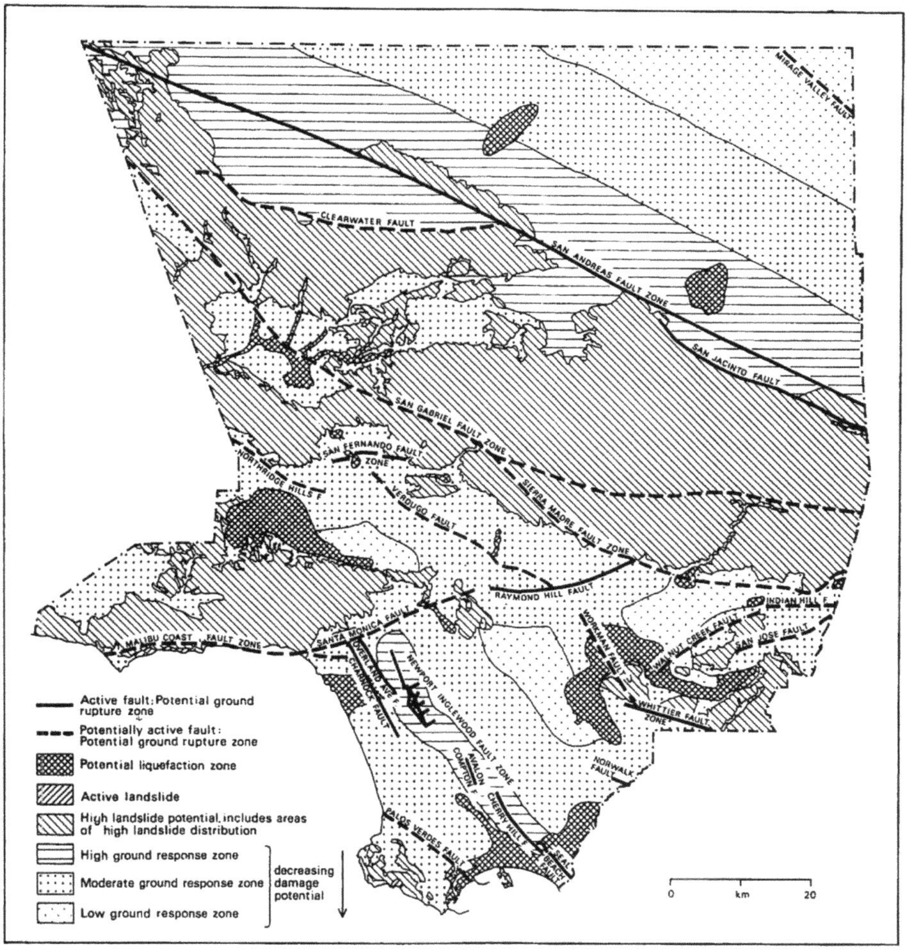

Figure 1.8 Seismic zoning map of Los Angeles County (after Los Angeles County, Department of Regional Planning 1974).

groundwater and seismic information. Active and potentially active faults, zones of potential liquefaction, active landslide areas, areas of high landslide potential, and three categories of ground response were identified (Fig. 1.8). Other consequences of this work included improved planning, zoning and building code legislation at state, county and city levels (e.g. State of California 1971a & b, Los Angeles County Board of Supervisors 1972, City of Los Angeles, 1978).

CLIMATE, WEATHER AND GEOMORPHOLOGY

The climatic scene The climate of Los Angeles (Bailey 1954 & 1966, US Weather Bureau 1966) is one of warm, dry summers, and cool, wet winters. Within this general context, regional topography imposes altitudinal gradients on temperature, relative humidity and precipitation, and significantly controls the movement of weather systems.

Spatial variations within the typically 'Mediterranean' seasonality of climate (Fig. 1.9) have been analysed by Bailey (1954, 1966) and others. In summer, the terrain regulates the in-blowing sea breezes, and where the maritime air is blocked by mountains, temperatures rise rapidly inland; as a result, a temperature inversion is common up to about 1500 m, above which temperature falls normally with altitude. In winter, the mountains block the inflow of low-pressure, rain-bringing systems, so that the west-facing slopes of the mountains are the wettest; interior basins tend to be drier and sunnier, and there are fairly strong 'rainshadow' effects (LACFCD 1963). In addition, offshore air movements, from the interior desert westwards, are important components of winter weather. Altitudinal control on temperature is stronger, and temperature contrasts are also greater in winter. The major weather types affecting the Los Angeles area are shown in Figure 1.9.

The topographically controlled climatic variations give rise to distinct climatic regions (Fig. 1.9). The coastal and marine climates are equable (with mean monthly temperatures in the range 10°–21°C), relatively dry (precipitation about 381 mm per annum), seasonally foggy and fairly well ventilated. In contrast, the interior valleys, and their transitional slopes to the mountains, are warmer and sunnier (with temperatures often over 26°C in

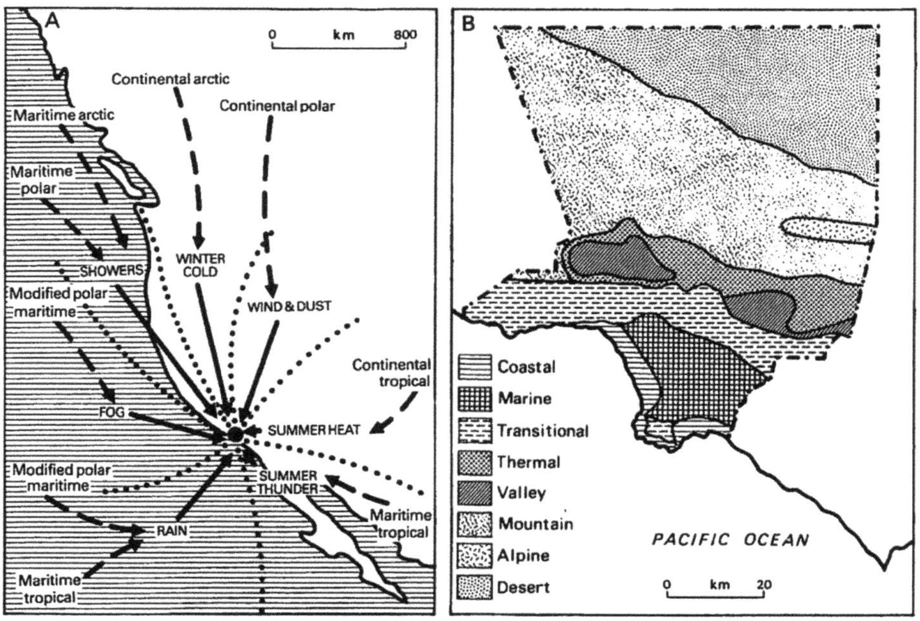

Figure 1.9 (A) Weather types of the Los Angeles region (after Bailey 1966). (B) Climatic regions of Los Angeles County (after Nelson & Clark 1976).

winter), but with a tendency for winter frosts on higher ground, and seasonally strong temperature inversions accompanied by high atmospheric pollution. Rainfall is also higher, with averages exceeding 508 mm per annum. The mountain areas are cooler (mean monthly temperatures not exceeding 21°C) with generally lower humidities; precipitation increases altitudinally, with totals annually of over 500 mm, often falling as snow on the highest ground. The desert (which in Los Angeles County is at high altitudes) lies in the rainshadow of winter storms, and has mean January temperatures of about 10°C with some frost, while in summer the July mean is about 32°C and thunderstorms may occur.

Geomorphologically, the two most important consequences of this climatic pattern lie in the control of the vegetation pattern (see Fig. 1.6) and hence of the pattern of erosion, and in the fact that precipitation is usually highest in the mountains, with the result that most geomorphological energy is concentrated there, and most fluvial discharge originates there before issuing into the more highly evaporative environment of the plains.

Climatic sequences and geomorphology Beyond the seasonal pattern of climate, longer-term fluctuations (for example in periods of storminess and drought) are important in determining the rates of geomorphological activity. Analysis of the climatic record – which extends back to 1769 – led Lynch (1931, 1948) to conclude that there has been no significant change in the mean climatic conditions in southern California over the last 200 years. Statistical analyses of the climatic record since 1850 by Cooke and Reeves (1976) confirmed this view for the period of reliable records; in addition, they showed that aspects of precipitation of probable relevance to runoff and erosion – for instance seasonal distribution, and intensity (mean daily, maximum 24-hour, and maximum daily) – also show no statistically significant secular trends since 1850.

Nevertheless, periods of relatively wet and relatively dry years are discernible. Table 1.5 shows these periods together with the major episodes in which flood and slope damage was reported to have been serious. From these data it is clear that floods are closely associated with the major wet seasons, when precipitation is well above normal (Lynch 1931). In addition, very severe floods often arise when a wet period terminates a series of dry years. In these circumstances, heavy rain storms may attack dry, vegetation-depleted surfaces – a good recipe for serious erosion and flooding. Such was the case in 1877–78, 1883–84, 1913–14, 1934 and 1978. The correlation between wet years and storm damage is close but not perfect, because damage depends on urban development and the effectiveness of hazard control (both of which change over time), as well as on the magnitude of the physical forces.

Some have identified a regularity in the occurrence of periods of wet years and periods of dry years. The annual rainfall figures for Los Angeles (Fig. 1.10), for example, might be deemed to have a general cyclicity about them.

Table 1.5 Summary history of storms, floods and slope failures, Los Angeles County.

Major slope-failure episodes	Flood years	Wet years and periods		Drought years and periods	
		Wet periods[a]	Wet years[a]	Dry years[a]	Dry periods[a]
	1770[g] 1825[c,h] 1833[c] 1840[e,h]		1810–21[g]	1822–32[g]	
	1850[e]		1849–50	1850–51	
			1852–53		
	1862[c,h]		1861–62		
				1862–63 1863–64	Summer 1862–Fall 1864
	1867[c] 1868[e]	Fall 1866–Spring 1868	1866–67 1867–68		
				1869–70 1870–71	Summer 1869–Fall 1871
				1872–73	
	1876[e] 1879[d]		1875–76 1877–78[b]	1876–77	
	1884[c,h]		1883–84[b]		Summer 1881–Fall 1883
	1886[c]/1887[d]		1885–86		
	1889[c,h]/1890[d]		1889–90		
	1891[d]			1893–94	
				1897–98	Summer 1897–Fall 1900
	1905[d] 1906[d]	Fall 1904–Spring 1907	1906–07		
	1909[d]		1908–09		
	1911[d]		1910–11		
	1914[d,h]/1916[e,h]	Fall 1913–Spring 1916	1913–14[b]	1912–13	

Table 1.5—*cont.*

Major slope-failure episodes	Flood years	Wet years and periods		Drought years and periods	
		Wet periods[a]	Wet years[a]	Dry years[a]	Dry periods[a]
	1921[e]			1923–24	Summer 1928– Fall 1931
	1934[e,h]				Summer 1932– Fall 1934
	1938[h]	Fall 1936– Spring 1938	1936–37 1937–38		
			1940–41		
	1943				Summer 1946– Fall 1950
1951/1952[f]			1951–52		
1956			1957–58	1958–59 1960–61	Summer 1958– Fall 1961
1962[f]				1963–64	
1965[f]			1966–67		
1969	1969				
				1976–77	
1978	1978				

Notes:

[a]Wet years are those that received more than 127 mm above the long-term mean; wet periods consist of two or more consecutive years in which precipitation exceeded the mean by 76 mm. 'Dry years' and 'dry periods' are defined by the same arbitrary amounts below the mean. The data used in this table relate to 12 recording stations in southern California, not to the Los Angeles City Hall records. After Cooke and Reeves (1976).

[b]Drought year or period terminated by a wet year. Horizontal lines indicate intervening years with approximately 'normal' precipitation. For other definitions, see below.

Data sources for floods in the Los Angeles River catchment:

[c]Los Angeles County Board of Supervisors 1915 (Superior Court of Los Angeles 1897, Daneri vs S. C. Railway Co.).

[d]Los Angeles County Board of Supervisors 1915 (US Weather Bureau records).

[e]US Army Corps of Engineers (1938). *Flood control in the Los Angeles County drainage area* (unpublished report).

[f]Jahns (1969).

[g]Lynch (1931).

[h]Floods reported by one or more of the above sources as causing unusually serious damage.

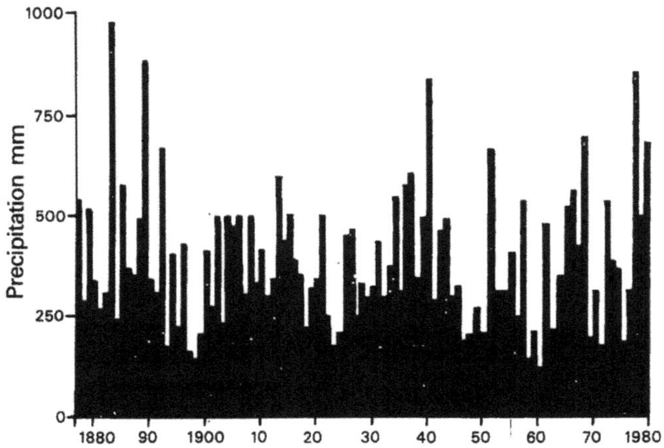

Figure 1.10 Annual rainfall in downtown Los Angeles (data source, Metropolitan Water District).

Bailey (1966, pp. 71–2) stated that:

> flood-making rains have decreased in frequency and severity in recent years. In the 20 years between 1944 and 1963 precipitation has only once exceeded 20 inches [508 mm] in Los Angeles (1952 with 24.95 inches), whereas in the first half of the precipitation record (1878–1920) totals at Los Angeles exceeded 20 inches in 11 out of 43 years.

However, this relatively dry period was not more significant than those from 1758 to 1810 (Lynch 1931), and from 1923 to 1934, or drier than the serious drought of 1976–77 (State of California, Office of the Governor 1977a & b). But it was a trough, and (like other troughs) certainly slowed the pace of geomorphological events, as revealed by the contemporaneous regional reduction of sediment yield (Ruby 1973). The wet periods – such as those from 1935 to 1944, and from 1964 to 1970 – similarly marked an acceleration of geomorphological activity. The notion of cyclicity gains support from the tree-ring studies of Schulman (1956), who suggested that sequences of dry years in southern California average about 15 years in length, and wet sequences average about 12 years, with a wet-and-dry cycle averaging some 27 years; Leighton (1966) recognised in the precipitation data three wet and three dry sequences since 1880.

Climatic crises: storms and fire weather Two weather situations are of fundamental importance to the dynamics of geomorphological processes in the shorter term, one direct, the other indirect: winter storms and the precipitation progression in winter, and 'fire weather'.

WINTER STORMS Storms in the Los Angeles area are primarily associated with winter cyclones originating in the north Pacific and moving SSE, and (relatively less important) with storms moving in from the south-west. Major storms may last for several days, especially if they become trapped in the Los Angeles Basin. The geomorphological effectiveness of storms is in general positively correlated with their precipitation amount, intensity and duration, and the quantity of precipitation prior to storms. However, critical thresholds must be crossed before saturation is achieved and processes such as erosion and slope failure are initiated. In a regional analysis, Cleveland (1971a) suggested that, for most types of surface, between 150 mm and 380 mm are required to reach saturation, with 254 mm being a reasonable average threshold value. Cleveland's map of the areas in California that are likely to receive 254 mm or more precipitation during a four-day storm every 25 years includes most of Los Angeles County south of the major divide of the San Gabriel Mountains. In some years (such as 1969) the threshold is exceeded throughout the area. Precipitation distribution during the wet season is as important as total precipitation because major storms, which usually occur in January and February, are likely to contribute proportionately more to runoff if they are preceded by above-average precipitation in October, November and December which tends to saturate the ground. Therefore, magnitude of runoff depends in part on the timing of storms within the rainy season. Nearly all major floods have followed an immediately antecedent precipitation of at least 200 mm. Conversely, heavy storms without adequate antecedent precipitation may cause little flooding. Such was the case in 1943, which had some of the most intense storms on record.

Just as the seasonal distribution of precipitation with respect to storms is geomorphologically important, so too is the distribution of rainfall intensity within a storm. This is well illustrated by Campbell (1975), who showed that the majority of soil slips in the Santa Monica Mountains during 1969 occurred in the two periods of maximum rainfall intensity following sufficient antecedent rainfall (see p. 44). No doubt a similar relationship holds for surface erosion. Larger slope failures (landslides) may not coincide with intense rains because there are often substantial time lags involved. Some extremely high intensities have been recorded: for example, over 16.5 mm fell in a minute at Opid's Camp in the San Gabriel Mountains in 1926 (a world record), and on 22 January 1943, 663.7 mm fell in a 24-hour period at Camp Le Roy in the same mountains (Sinclair 1954).

'*Fire weather*' is very different from storm weather. It is of two main types (Bailey 1966). First, lightning is a significant natural cause of fires. Lightning weather tends to increase in frequency with altitude and distance inland and occurs mainly in the mountain areas. About 40% of the days with thunder in the Angeles National Forest are associated with records of fire, suggesting a strong link between lightning and fire; and the rain from summer thunderstorms may be inadequate to extinguish the fires (Bailey 1966). Of the fires in the Los Angeles River watershed area of the Angeles National Forest between 1939 and 1959, 10.2% were probably initiated by lightning (USDAFS 1961).

Secondly, fires are especially associated with 'Santa Ana' wind conditions, which can arise from several synoptic situations (e.g. Schroeder 1964, Hull *et al.* 1966, Weide 1968). In these conditions, föehn-type winds blow from the north and north-east over the mountains from the desert. They are initially warm and dry, but as they warm adiabatically during their rapid descent down valleys into the Los Angeles Basin, their relative humidities may fall to as low as 2%, and their temperatures rise to over 37°C. Such winds are commonest and most effective in the autumn, when they blow over and desiccate further vegetation already tinder-dried in the preceding long, dry summer (e.g. Edinger *et al.* 1964, Weide 1968, Countryman 1974). All that is required is a spark, usually supplied by man (USDAFS 1961). The winds (sometimes blowing at over 120 kph) supply oxygen to the fire, bend flames forward, and scatter firebrands ahead of the main fire front (e.g. USDA 1957, Ryan *et al.* 1971).

An analysis of the elements of fire-promoting weather (such as summer drought and, perhaps, drought in the preceding years, expressed in terms of fuel moisture, low relative humidity, lightning risk, high air temperatures, and high wind velocities) is normally incorporated into fire-hazard prediction indices, examples of which include the Fire Weather Severity Index, the Fire Danger Rating Index and the Fire Load Index (e.g. Weide 1968, Countryman *et al.* 1969, Deeming *et al.* 1972). Although weather details associated with fires and Santa Ana wind conditions need no closer examination here, it should be noted that the spread of fires and the ultimate area of devastation often reflect the strong topographic control of wind movement (e.g. Fosberg 1965), and the fact that wind directions may change dramatically (even be reversed) during the day at ground level as a result of the changing surface pre-eminence of Santa Ana and onshore winds (e.g. Fosberg *et al.* 1966, Weide 1968, Countryman *et al.* 1969, Ryan *et al.* 1971).

FIRE

Geomorphological effects of fire

FIRE AND ACCELERATED EROSION Burning of vegetation has far-reaching consequences for geomorphological activity. Of the greatest importance is the effect of fire on erosion and sediment yield, which can conveniently be summarised in terms of a 'fire-induced sediment cycle' (Smalley 1971, Smalley & Cappa 1971). The cycle begins with a fire (Fig. 1.11) that destroys the shrub canopy and most trees, grass and grass-root net and litter, and converts humus into ash. Clays may be calcined and flocculated, and infiltration rates may be reduced as a result of structural changes in the soil and burned-litter layer that lead to increased water repellency (p. 9). Thus, the ability of the slope to store water and to lose water by transpiration is reduced,

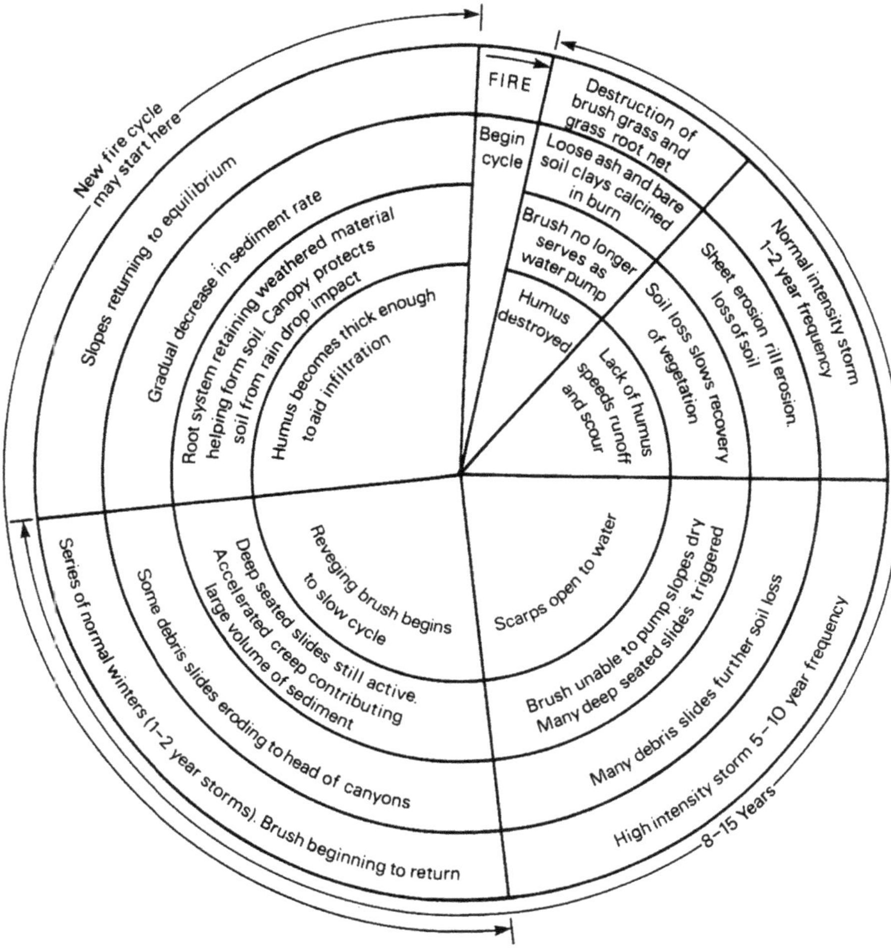

Figure 1.11 The 'fire-induced sediment cycle': a summary of common consequences of a fire in southern California mountains (after Smalley & Cappa 1971).

movement of water into the soil is restricted, and raindrops more commonly strike the soil surface.

All such changes tend to increase surface erosion and sediment yield. The fire effect on erosion is shown by studies in the San Gabriel Mountains following a fire in 1959 (Fig. 1.12, Krammes 1960, 1963, Doehring 1968). Here, in the years of monitoring preceding the fire, both wet and dry season erosion rates were relatively slow, but much faster on south-facing 'rejuvenated' (undercut) slopes. Acceleration of erosion actually began during the fires as a result of stampeding wildlife and mobile burning debris (Doehring 1968). After the fire, wet and dry season movement accelerated (even though the rainfall relevant to wet erosion was only 70% of normal), and the acceleration was most pronounced in the dry season. The post-fire acceleration was spectacular in the first post-fire year, with annual soil loss increasing 10–16 times; thereafter,

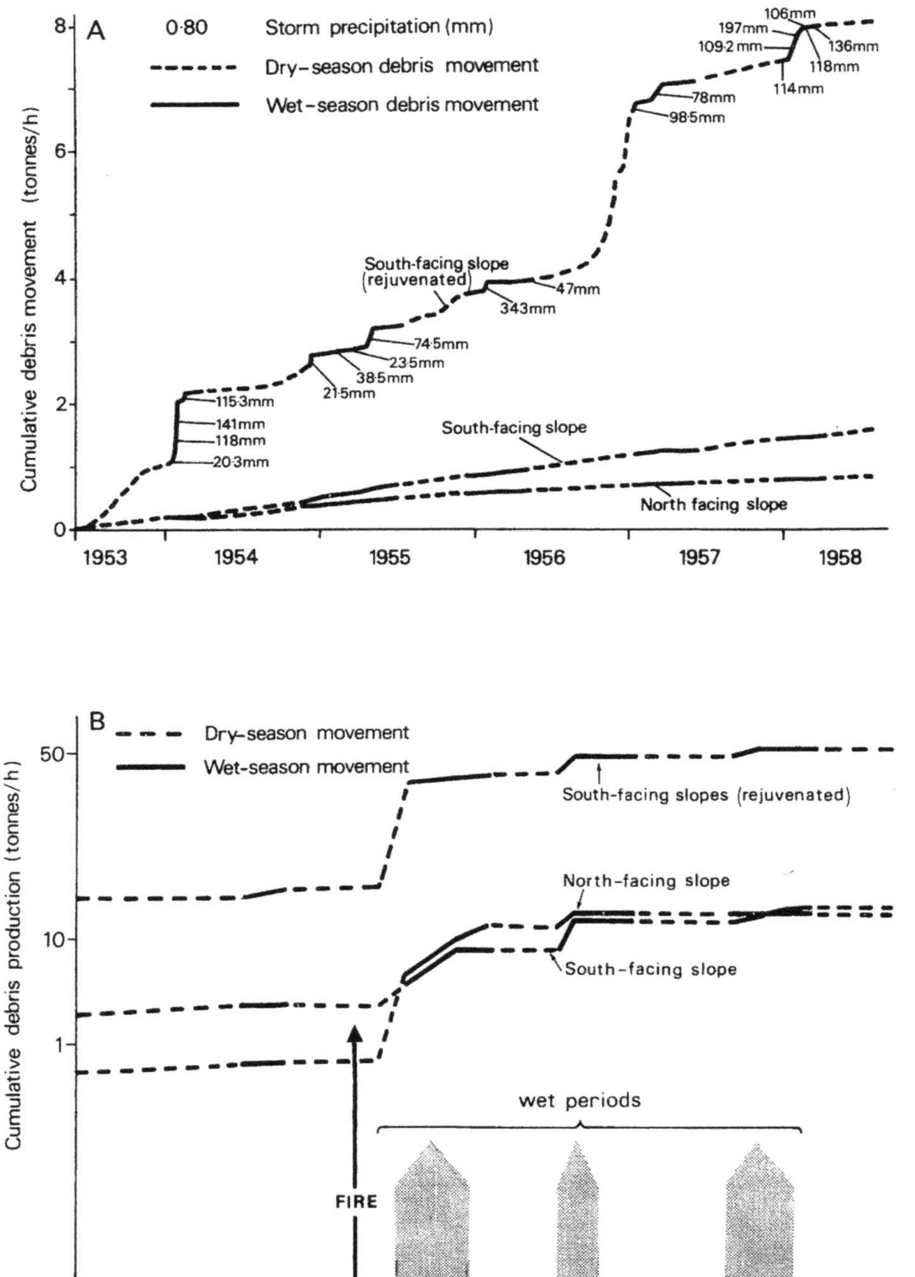

Figure 1.12 Seasonal debris movement before and after a fire at debris-monitoring sites on slopes in Wilson diorite in the San Gabriel Mountains. (Part (A) after Anderson *et al*. 1959, part (B) after Krammes 1960 & 1963).

deceleration was evident, although high yields were associated with intense storms.

FIRE AND SLOPE FAILURE Fire-generated changes may also encourage slope failure in the form of soil slips and landslides, although here the effects are less clear. For example, Rice (1973) compared 'landslide erosion' (essentially soil slips) on three similar chaparral areas after the storms of 1969; one area had burned the previous summer and had slope failure over 10.0 m^3/ha; another area burned nine years previously lost 297.1 m^3/ha; and an area unburned for 50 years lost 16.04 m^3/ha. Rice suggested that the main reasons for the low rate of slope failure in the recently burned watershed appear to be the development of the water-repelling layer following the fires (which restricts increases in pore-water pressure and soil saturation), and the fact that root structures were still intact. In the 50-year-old unburned watershed, slope failure was higher because of high infiltration rates and despite high vegetation cover; in the 9-year-old watershed, damage was greatest because infiltration rates had probably been restored, but roots had decayed and new vegetation had not developed an effective new root structure. Soil slips may actually be promoted by vegetation regeneration (especially chaparral) because infiltration is increased and thorough saturation of the soil mantle is more rapid (e.g. Campbell 1975). Transformation of chaparral to grass may also increase superficial slope failure (Corbett & Rice 1966, Rice 1973). Major landslides often experience a time lag before failing following major storms or similar events, so the effect of fire is even less certain on them, but Cleveland (1971a) and Smalley and Cappa (1971) judged the effect to be important.

FIRE SEQUENCES The precise geomorphological effects of fire depend very much on the nature of the surface, the time the fire occurs with respect to the rainy season, and the magnitude, frequency and duration of rainfall events in the years following the fire. The individuality of fire effects on erosion is graphically described in the San Dimas Experimental Forest after fires in 1938, 1953, and 1960 (USDAFS 1954, Doehring 1968). The most serious consequences arise when a heavy storm occurs immediately after a fire. This is a fairly common event, as fires tend to be concentrated in the Santa Ana conditions of autumn prior to the rainy season. In these circumstances, runoff, soil erosion by water, slope failure, and ultimately sediment yield rise rapidly, the last-mentioned to levels as much as 20–30 times higher than normal (e.g. Los Angeles County Flood Control District – LACFCD – 1959).

After some years, vegetation becomes re-established, and the slopes re-adjust to a new equilibrium, so that sediment yield falls back to a 'normal' level. Beyond that, much will then depend on subsequent fire history. The longer the interval between fires, the deeper is the soil layer available for erosion, the greater is the available quantity of fuel (and hence the hotter the subsequent fire), and the greater is the hydrophobicity (and thus the greater the

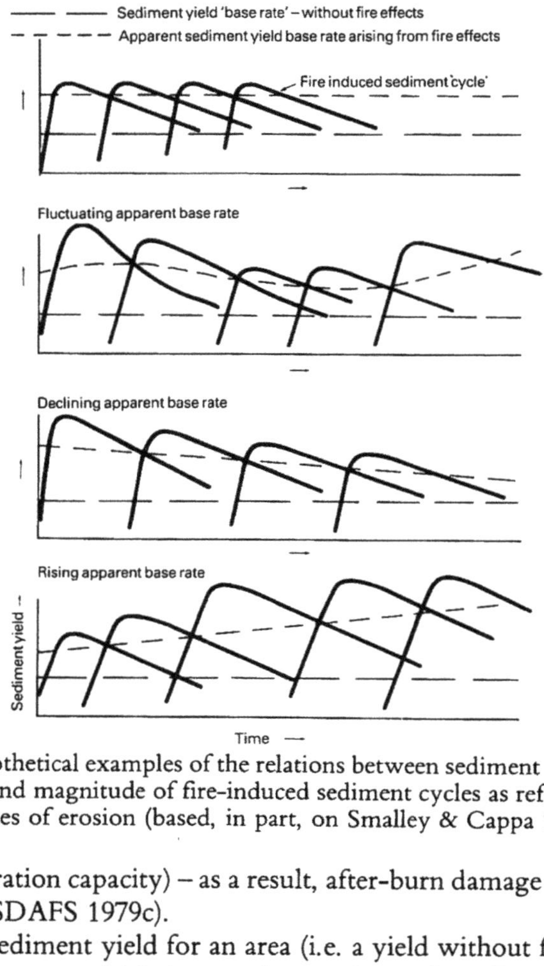

Figure 1.13 Hypothetical examples of the relations between sediment yield over time and the frequency and magnitude of fire-induced sediment cycles as reflected in trends in apparent base rates of erosion (based, in part, on Smalley & Cappa 1971).

reduction in infiltration capacity) – as a result, after-burn damage is likely to be increased (e.g. USDAFS 1979c).

A base rate of sediment yield for an area (i.e. a yield without fire influence) can be envisaged upon which is superimposed a series of peaks related to fire-induced sediment cycles (e.g. Smalley 1971). These peaks may be superimposed to give a false impression of sustained higher base rates, and, as Figure 1.13 shows, these apparent base rates may vary systematically over time according to the frequency and magnitude of fire-induced sediment cycles.

The nature and rate of vegetation recovery are fundamental to determining the shape of fire-generated sediment-yield curves. Hanes (1971), in an important ecological analysis, emphasised that the species composition and rates of chaparral succession to climax following a fire are not uniform, but are strongly influenced by aspect, coastal or desert exposure, and altitude. For example, he found that full recovery in coastal chaparral is slowest on south-facing slopes below 900 m, and fastest on north-facing slopes above 900 m; and the time required for regeneration to climax may exceed 30 years. Such observations lie behind the regional assumptions normally built into predictions of sediment yield. In this context, the empirical analysis of Rowe *et al.* (1954) is

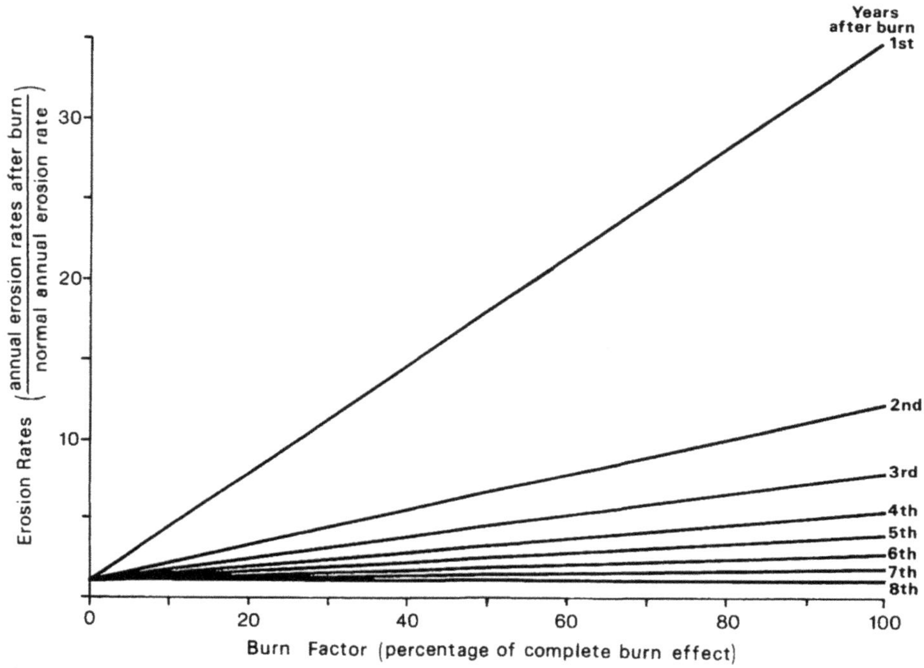

Figure 1.14 Relations between proportion of watershed burned and the ratio of annual erosion rate after a fire to normal annual erosion rate in the Los Angeles region of the San Gabriel Mountains (after Rowe *et al.* 1954).

fundamental, and has been used as a basis for most subsequent generalisations (e.g. LACFCD 1959 & 1979b, Tatum 1963). They showed that in the San Gabriel Mountains normal base-load erosion was increased in the first year following a fire by 15–35 times, the amount of increase depending on various factors, especially aspect and location. Equally importantly, they suggested that the fire effect on erosion rates lasts with diminishing influence for approximately a decade and the effect is related, *inter alia*, to the extent a watershed is burned (Fig. 1.14). The rise and fall of erosion rates after a fire are partly a function of changing vegetative protection; they may also be influenced by a decline in sediment availability. Fire effects and vegetation recovery rates vary slightly according to the location within the mountains (e.g. LACFCD 1959, Appendix E).

The spatial and temporal contexts of fire in Los Angeles County

SPATIAL FREQUENCIES The fire history of Los Angeles County (summarised in Fig. 1.15) is based on the analysis of fire records in the library of the LACFCD. The comprehensiveness of these records is not known, but they appear to be slightly fuller than those used by Weide (1968) in the Los Angeles County Fire Department. Although they include fires from 1878, few are recorded before 1900, and it seems likely that the information becomes

Figure 1.15 Fire frequency in the mountains of Los Angeles County, 1878–1963 (based on an analysis of fire records in the LACFCD).

more accurate after 1915, when the LACFCD was created. These records, however accurate, form the basis of the following discussion.

Given the importance of fire in influencing sediment yield and other geomorphological processes, it is important to determine in Los Angeles County the spatial pattern of fire frequency, any frequency–magnitude changes that may have occurred, and the timing of fires with respect to seasonal storms.

Figure 1.15 shows that fire frequency is spatially variable, with the frequency being highest along the coastal front of the Santa Monica Mountains and the southern front of the San Gabriel Mountains. Here, some areas have been burned three or four times between 1910 and 1963, giving a recurrence interval of 17.6–13.25 years – a frequency, that is, of the same order of magnitude as the time normally assumed to be required for vegetative recovery following a fire. Such areas might be expected to generate high sediment yields. The relatively high frequency of fires on the southern fronts of the Santa Monica and San Gabriel mountains is also apparent from Kenyon and Jones' (1959) study of fires between 1919 and 1958. They showed that the total burned area in the 'fire areas' behind Santa Monica and Pasadena was likely to be equivalent to that of the whole 'fire area' in 41–3 years, whereas in the 'fire area' on the northern slopes of the San Gabriel Mountains the period was as high as 120 years. In a 'fire frequency damage probability study' for the 'front' (i.e. south-facing catchments) and 'back' country of the San Gabriel Mountains, the USDAFS (1979b) demonstrated (using similar data to those used in Fig. 1.15) that the probability of a fire exceeding 16.9 ha in any one year at any site is 1.52% for the frontal areas and 0.6% for the 'back' country.

The reason for the relatively high frequency of fires on the southern flanks of the San Gabriel and Santa Monica mountains lies mainly in the increased opportunities for man-induced fires in these areas. Whereas some fires have been started by lightning (which may occur in poor burn conditions if it is associated with heavy rain), and some fires are set deliberately, most fires are caused accidentally by man. Some are associated with powerlines, but camp fires, matches and sparks from vehicles (all associated with visitors) are pre-eminent causes – and visitors concentrate in these south-facing mountain areas. The USDAFS (1973a) estimated, for example, that 52.6% of man-caused fires – accounting for 54.7% of the area burned within the Los Angeles River Watershed – originate within 1.6 km of the southern national forest boundary along the foothills.

CHANGES OVER TIME Table 1.6 shows that the frequency of both small and large fires (i.e. under 400 ha and over 400 ha) has increased over time. The average size of fires has also increased. Although these trends could in part reflect improved data collection, they almost certainly also represent the net result of two mutually opposed trends: a rise in the frequency of fire starts arising from increased use of mountain areas by people on the one hand (Table 1.7), and somewhat slower improvement of fire prevention and control on the other. Control and prevention are not uniformly applied or effective, for they

Table 1.6 Relations between fire size and percentage fire frequency, 1878–1960 (from the fire records in the LACFCD).

	Fire frequency (% of total recorded fires)								
Fire size (ha)	1870–80	1880–90	1890–1900	1900–1910	1910–20	1920–30	1930–40	1940–50	1950–60
>400	0.2	0	1.6	3.5	11.0	20.2	12.6	22.4	28.6 (n = 108)
<400	0.9	0	2.8	1.8	5.5	18.0	12.0	24.0	34.0 (n = 370)

Table 1.7 Visitors to the Angeles National Forest (USDAFS 1973a).

Year	Number of visitors
1939	1 610 000
1949	1 820 000
1959	3 140 000
1969	6 980 000

appear to be most successful within the areas covered by the 1941 and 1960 fire protection programmes for the Los Angeles River Watershed (USDAFS 1973a).

GEOMORPHOLOGICAL IMPLICATIONS. Spatial concentration of fires and change of fire frequency have important geomorphological implications. In essence, they mean that the accelerated sediment yields from the southern flanks of the Santa Monica and San Gabriel mountains are at least likely to be sustained. It is possible, too, that as population and visitor-use patterns change, these most vulnerable zones may be extended, perhaps to include such areas as the hills surrounding the north and west ends of San Fernando Valley, and the Santa Clara Valley.

In the shorter term, the timing of fires in relation to the storm season is important. In general, it seems probable that the later a fire occurs in the autumn, the less time there will be for vegetation (especially the grasses) to become established before the storm season and, as a result, the greater will be the erosive effect of the storms. Figure 1.16 shows that whereas the number of fires is greatest in July, the number of fires over 400 ha is greatest in September and, most significantly, a quarter of all fires over 4000 ha occur even later, in October. The principal reason for this is that the larger fires tend to occur later in the autumn at the time of severe Santa Ana conditions. The main consequence is that large areas are exposed to the attack of winter storms before vegetative recovery and protection are possible.

Geomorphological activity will be greatest when areas burned before the rainy season are struck by particularly severe storms. The coincidence of these events cannot be predicted, but in the past many major episodes of slope and channel damage accompanied such spatial and temporal coincidences (see Part 2).

Management of fire and its geomorphological consequences Wildland fires in Los Angeles County, as elsewhere in California (Phillips 1971), are a major problem of environmental management because of the threat they present to life and property, and the ways in which they can exacerbate the flooding and erosion hazards. Their management in Los Angeles County is largely the responsibility of the County Fire Department (with which most cities contract), certain cities, and the USDAFS.

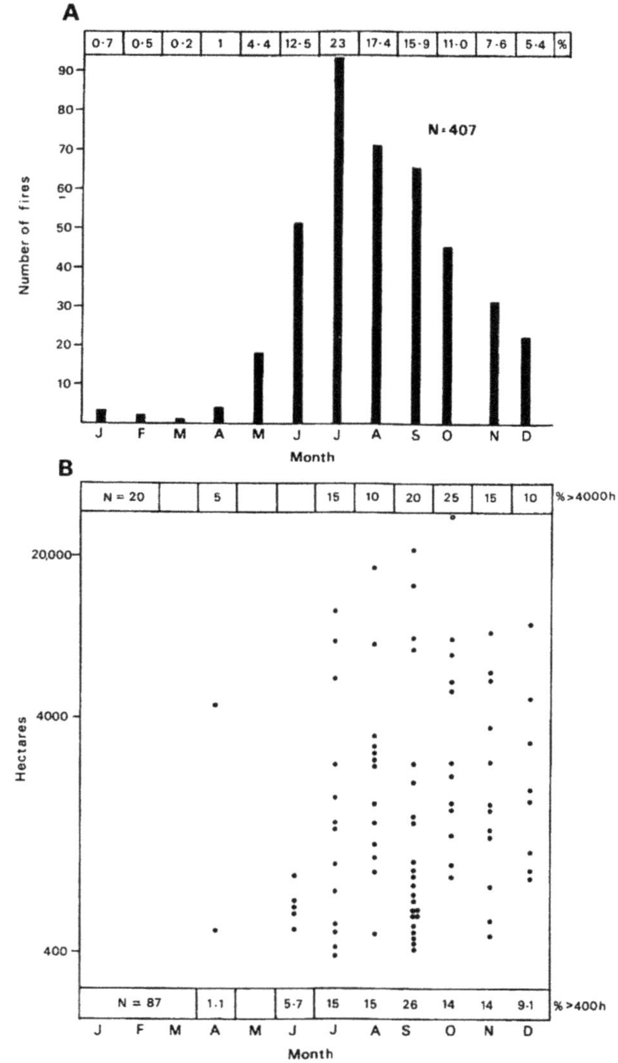

Figure 1.16 Relations between frequency (A) and size (B) of fires, and time of year (based on an analysis of fire records in the LACFCD).

Several efforts to manage fires have an effect on geomorphological processes, directly or indirectly. Only such efforts are discussed here; for a more comprehensive account of fire management, see State of California Department of Conservation (1972). The main areas of management include forecasting, suppression techniques, rehabilitation, and a whole range of preventative measures (e.g. State of California, Department of Conservation 1972, Countryman, 1974). Forecasting involves prediction using a group of climatic indices (see p. 24); suppression techniques – which include a range of control measures from brush-beating to aerial bombardment – are integrated into fire-fighting procedures (e.g. Los Angeles County Fire Department 1967a). Successful forecasting and fire suppression serve to reduce the area affected by fires and thus, it is hoped, restrict their geomorphological consequences.

REHABILITATION AND PREVENTION Of greater importance are the rehabilitation and preventative techniques, especially because the management aim of reducing fires may not be wholly compatible with the less visible objective of reducing erosion and flooding. The main techniques involve local authority ordinances directed towards fire prevention, efforts within the 'watershed fire areas' to reduce the energy output of fires through management, and attempts to reduce the hazard through control of fire spreading and improving access to, and suppression, of fires once they have begun.

Local authority controls are incorporated in city and county fire codes. For example, Article 27 of the County of Los Angeles Fire Code (e.g. Los Angeles County Fire Department 1967b) includes restrictions on open fires and smoking (section 27.300), detailed controls for brush clearance and vegetative growth, and limitations on activities in hazardous areas (section 27.400). Of particular interest is the requirement to clear vegetation along powerlines (section 27.300) and from around buildings (section 27.301). The City of Los Angeles has adopted a similar brush-clearance ordinance (136195) in the 'mountain fire district' which requires the maintenance of firebreaks around buildings by reducing brush height and removing branching growth and leafy foliage (City of Los Angeles 1968). Together with the extensive tracery of firebreaks and access roads in the mountains (USDAFS 1979c), the vegetation reduction and clearance required by such ordinances means, in effect, that large areas are 'bare' of vegetation, or are only lightly covered, and are directly exposed to storm attack. Thus fire and fire damage may be reduced, but erosion from the protected areas may be unintentionally increased. For example, as long ago as 1938, extensive damage by storms to trails and firebreaks was recorded (Los Angeles County, Department of Forester and Fire Warden 1938). Now, as then, many mountain roads are severely eroded in storms (USDAFS 1979a & c).

A second possible conflict between management objectives arises in the field of vegetation rehabilitation and modification. Vegetation rehabilitation involves, most commonly, emergency reseeding of burned areas with fast-germinating grass seed before winter rains fall (Phillips 1971). The aim of reseeding, of course, is to prevent soil erosion. But (as discussed above) the transformation of chaparral to grass can enhance slope failure in the form of soil slips because the binding influence of root structure has been depleted. The creation of fuel breaks may have the same effect (Rice *et al.* 1982). An alternative strategy of great potential value involves the modification of fire fuel by controlled rotational ('prescribed') burning, which creates a mosaic of age classes of vegetation and reduces fire susceptibility, because once burned, vegetation does not burn again for ten years or so (Countryman 1974). If such a practice were to be widely adopted – and there are signs that it is favoured (USDAFS 1982) – its effect on geomorphological activity cannot be predicted with certainty. It seems possible that it will increase sediment yield if storms immediately follow burning, but the spatial mosaic of vegetation of different ages in a watershed might tend to reduce the peaks in the sediment-yield curves of unpremeditated fires. The rotational burning method, by reducing the fuel available to unpremeditated fires, may reduce the development of erosion-accelerating hydrophobicity (Rogers 1982). On the other hand, Rice *et al.* (1982) have suggested that on steep slopes where stability depends partly on roots, the practice reduces root biomass and may consequently increase soil slippage. The policy has the advantage of recognising the fundamental rôle of fire in the evolution of chaparral vegetation (Hanes 1971). Fire has always been a major factor in chaparral development, and to use it as a management tool may be more sensible, cheaper and more successful than the earlier, continuing and expensive attempts to exclude fire altogether. The difference in policies is, of course, that one seeks to prevent and control accidental conflagrations, while the other seeks to create controlled conflagrations.

URBAN DEVELOPMENT AND GEOMORPHOLOGY

The growth of metropolitan Los Angeles in the 20th century has been spectacular, but data on the details of urban development are surprisingly difficult to find. One measure is the age of housing, as shown by Nelson and Clark (1976, Fig. 1.17). Another is based on the record of subdivisions in the annual planning reports of the City of Los Angeles up to 1959 (Fig. 1.17). A comparison of these patterns with the distribution of orchards and agricultural land in 1939 (Fig. 1.17) leaves no doubt that urban growth has been very largely at the expense of agriculture. The only important exception to this generalisation is the extension of residential property into the mountains, principally the flanks of the Santa Monica and San Gabriel mountains and intervening ranges (such as the Verdugo Hills).

The geomorphological consequences of urban development are both indirect and direct. Indirectly, the huge increase of population and wealth that has accompanied urban growth has increased pressures on the environment. For example, increased recreational use of the mountains is reflected in increased fire frequency which, in turn, leads to enhanced erosion. Also indirectly, the growing affluence of the community has provided the means to manage the environmental problems that have arisen.

Directly, urban growth is responsible for transforming geomorphological processes into community hazards. Most obviously, settlements have extended into areas where slope erosion, slope failure and flooding have always existed. Nowhere is this more apparent than on the alluvial fans flank-

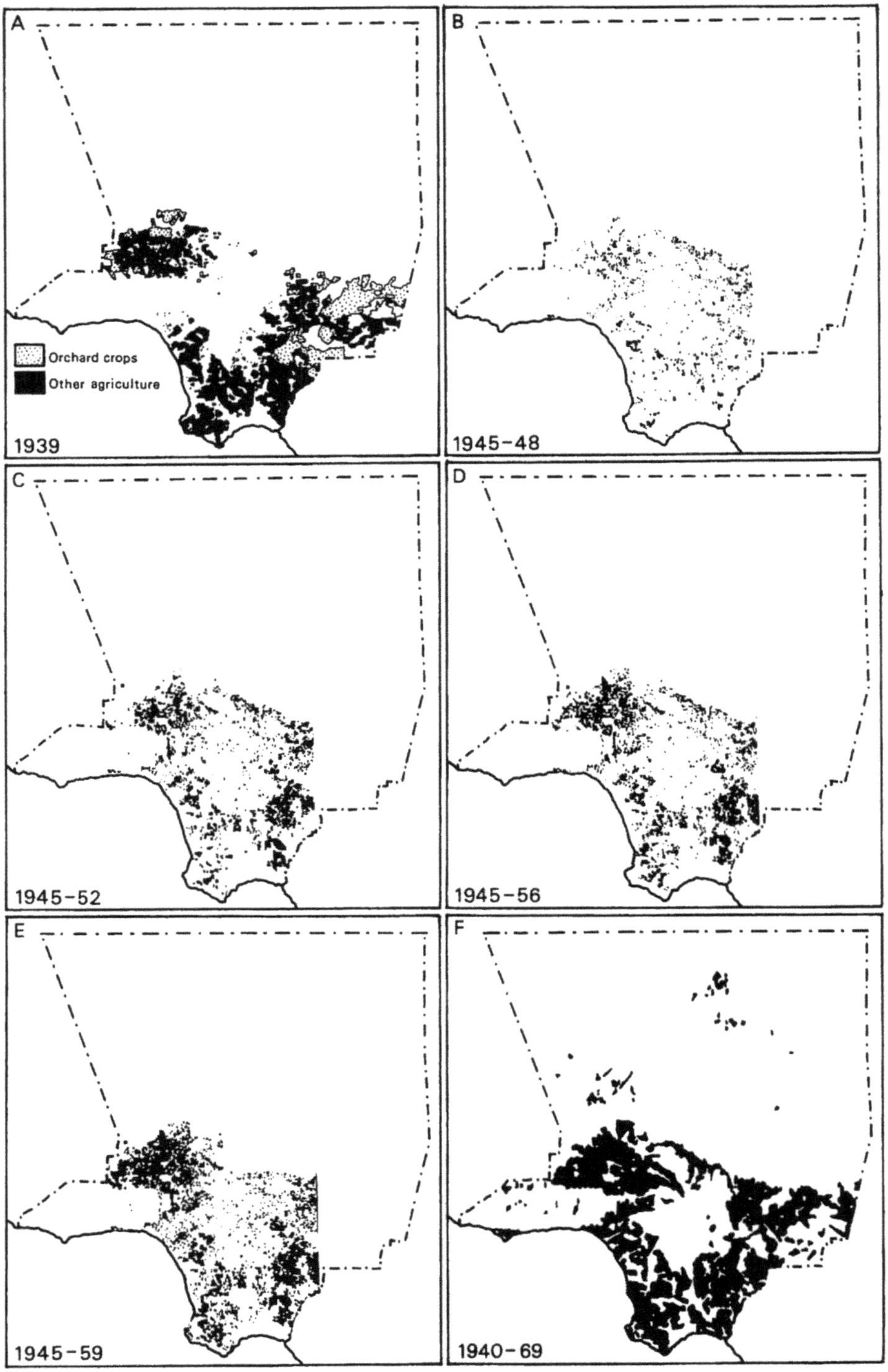
A
Orchard crops
Other agriculture
1939
B
1945-48
C
1945-52
D
1945-56
E
1945-59
F
1940-69

ing the San Gabriel Mountains, where the cities of Glendora, Azusa, Glendale, Pasadena, Sierra Madre and such communities as La Crescenta, Verdugo, Montrose and La Cañada figure prominently in the archives of storm damage. These settlements have become obstacles to water and sediment movement in the natural drainage systems. At the same time, urban growth has radically altered surface characteristics and hydrology. The transformation of an orchard or a field into a city, with its efficient artificial drainage system and its surfaces of relatively impermeable concrete, roofing materials and asphalt, profoundly influences the quantity and rate of runoff, tending to increase both. For example, the annual peak discharge of the Los Angeles River near the Wardlow gauge (Long Beach) as represented by 10-year running means, has risen very substantially since 1941 (Evelyn, 1982). The act of development itself, usually widespread at any one time, may denude large areas of vegetation and expose vulnerable soils to erosion, and, unless it is controlled, may temporarily stimulate sediment yield.

The development of hillside housing is a phenomenon of recent decades that reflects several contemporaneous trends. Firstly, as the availability of land for urban development on the plains has diminished and land values have risen, it has become more economically acceptable to develop hillsides and other 'marginal' land. Secondly, it has become increasingly fashionable to seek home sites in the hills, where the advantages of a view over sea or city are enhanced by the avoidance of the smog hazard. And partly as a result of these trends, the means of building safely on steep slopes have been developed: techniques of foundation engineering and earth moving have advanced rapidly. Hillside developments – like their counterparts on the alluvial plains – transform the ground surface and modify its hydrology. In addition, however, they have tended to exacerbate the natural instability of slopes. The reforming of slopes, mainly by 'cut-and-fill', may disturb slope equilibrium and, under certain conditions, promote slope failure. Furthermore, the innocent creation of lush vegetation in gardens and parks may cause problems, because in the semi-arid environment irrigation is essential; over-irrigation (a common error) can lead to unanticipated increases in soil moisture and pore-water pressures that may contribute to slope failure.

Like fire, urban development changes geomorphological systems in ways that are both spatially and temporally variable, and it thus adds significantly to the constantly changing nature of geomorphological hazards.

Figure 1.17 Settlement growth and land-use change in Los Angeles County. (A) The land-use pattern in 1939 (after Nelson & Clark 1976). (B) – (E) Urban growth from 1945 to 1959 (based on the record of subdivisions in the annual planning reports of the City of Los Angeles up to 1959 – no data for east and north of county). (F) Housing growth, 1940–69 (after Nelson & Clark 1976).

THE CHANGING EFFECTS OF HAZARD MANAGEMENT

Management responses to geomorphological hazards, ranging from land-use zoning and building design control to floodwater storage, are considered extensively in Part 3, but it is important to emphasise here that as a whole they too have substantially modified the spatial and temporal patterns of the problems. This effect is dramatically exemplified by the changing pattern of flooding in Los Angeles County which has occurred during the 20th century. The first well-documented flood (in 1914) involved serious flooding along the main waterways in the Los Angeles Plain near to the sea (see Fig. 2.1, p. 59). By 1931, the predicted area of inundation during a flood 50% greater than that in 1914 was substantially extended, especially in San Fernando Valley and the San Gabriel Plains, as a result of urban growth without flood control, but it had been reduced in the downstream areas (LACFCD 1931). Thereafter, federal, state and county funds having been secured for flood protection, massive engineering works radically reduced the area of potential flooding in urban areas. As a result, by 1970 the flood-hazard zones were confined to a few uncontrolled valleys in the south-facing mountain areas, and the largest remaining hazard area was in the desert region beyond the boundary of the original flood control district (Los Angeles County Board of Supervisors 1974).

Geomorphological processes

The major geomorphological processes identified in Figure 1.1 are reviewed here.

Slope decomposition

The preparation of surface debris for erosion, by rock weathering and soil formation, is probably the least studied aspect of the geomorphological system in southern California. That debris is available in large quantities is beyond doubt, but the processes responsible, their rates of operation, and the possibility that some weathering mantles are inherited from previous climatic conditions are largely matters of speculation. Reconnaissance observations suggest that the disaggregation of sedimentary rocks is influenced by wetting and drying of expansive clay minerals – such as montmorillonite – and by the direct impact of raindrops, as well as by other physical and chemical weathering processes. In the areas of crystalline rocks, chemical decay is clearly important – as the occurrence of deep weathering profiles reveals – and some of these weathering mantles may be inherited from a time in the past when climatic conditions were different (e.g. Cooke & Mason 1973). At higher altitudes, it is possible that frost action is important, and salt weathering may be significant in coastal and desert areas.

Slope erosion

Retzer *et al.*'s (1951) field observations suggested the following major locations of slope erosion in the mountains: stream banks and slopes over-steepened by faulting, slope failures, river undercutting or man-made excavations, south-facing slopes (which in general have thinner vegetation and soil cover), slopes above the angle of repose of loose material, and colluvial/alluvial deposits. It is quite common for steep slopes to be convex in profile (Scott 1971) and for the steeper, lower segments to be particularly active erosion sites. The smoother, less active slopes of the foothills and isolated upland areas are

DOI: 10.4324/9781003168645-3

primarily eroded near their bases (usually by gullies), where surface runoff and soil moisture are greatest. In his study of the Verdugo and San Rafael hills, Strahler (1950) noted that slopes protected from recent basal erosion have significantly lower angles, and he concluded that these slopes declined in angle 'when only sheet-runoff, creep and other mass-wasting processes operated on the slope' (p. 813).

The two major processes of slope erosion are dry erosion and wet erosion.

DRY EROSION

Dry erosion (sometimes called 'dry ravel') is a distinctive process in the semi-arid mountainous environment of southern California, where slopes are steep, and dry, loose material litters the surface. Movement is mainly by sliding and is primarily controlled by gravity. It is greatest on slopes above the angle of repose of unconsolidated material. The threshold slope angle is about 70%, and many valley-side slopes in the San Gabriel Mountains equal or exceed this figure (see p. 5). Movement includes the full range of debris from boulders to clay particles; it is restricted by vegetation (and, consequently, accelerated by fire), and may be triggered by wetting-and-drying, earthquakes, wind or animals.

WET EROSION

Wet erosion involves raindrop-impact erosion and runoff erosion. The former is not thought to be very important on vegetated surfaces (Rice 1973), but where the vegetal cover is removed (especially by fire), it can be very effective. The dynamics of raindrop erosion are well understood. Its effectiveness is in general proportional to the kinetic energy of storm events and slope inclination, and inversely proportional to the resistance of surface materials. Runoff erosion has also been extensively studied (Kirkby & Morgan 1980, Morgan 1980), although not in Los Angeles County. The principal mechanism depends on the generation of surface runoff when the rate of precipitation exceeds the rate at which water can infiltrate the soil. The effect of runoff is proportional to the ratio of forces applied to the resistance of surface material (including vegetation). The forces are proportional to the volume and velocity of flow which, in turn, depends on the rainfall intensity–infiltration capacity ratio (supply rate), slope steepness, and surface roughness. Soil resistance is influenced by the nature of soil material, vegetation cover, and root network. Fire is important because it not only removes vegetation, thus increasing raindrop impact erosion, but it also increases the hydrophobicity of soil, thereby increasing runoff and runoff erosion. Rilling by water erosion is common on fire-denuded surfaces, and may arise from the 'increase of positive pore-water pressure and reduction of effective normal stress in the saturated surface layer developed above the hydrophobic layer' (Wells & Brown, quoted in Howard 1982, p. 407).

There is a strong seasonality about wet and dry erosion, as studies at sites on

Wilson diorite in the San Gabriel Mountains have shown (Anderson *et al.* 1959, Krammes 1963). Figure 1.12 (see p. 26) reveals a distinct seasonal pattern on south-facing slopes in which dry erosion during the dry season was relatively continuous and produced the base-load and, indeed, the majority of debris erosion. The first rains reduced erosion by providing increased cohesion to the dry surface; then runoff surges accelerated erosion rates. Anderson *et al.* (1959) estimated that even under wet conditions on the most vulnerable sites, a rainfall intensity of over 6.35 mm/h was necessary to produce such surges. In general, wet erosion was directly correlated with rainfall intensity. Dry erosion was at times reactivated in prolonged dry spells between winter storms. Both dry and wet erosion were accelerated by fire (see Fig. 1.12, p. 26). Overall, recorded average annual erosion rates at the study sites ranged from 0.44 t/ha at a non-rejuvenated, north-east-facing site to 7.97 t/ha at a rejuvenated, south-east-facing site.

Slope failure

Slope failures have been classified in various ways using different criteria, but three types are commonly recognised (e.g. Varnes 1958, Morton & Streitz 1967): falls (movement of debris through the air), slides (downslope movement along slip or shear surfaces), and flows (downslope displacement of material as a viscous fluid). In southern California, the major phenomena are slides, especially soil slips and landslides; less important are rockfalls and subsidence due to groundwater and oil extraction. Debris flows are also fundamental, and are considered below in the context of channel activity, although some slips and slides merge downslope into debris flows. Indeed, the distinction among the features is somewhat arbitrary, for a continuum exists and composite features are common. Storm-related slope failures are often collectively referred to locally as 'mudslides'. The different types of slope failure are found chiefly in the mountain areas, in the intermediate hilly uplands (such as the Puente and Verdugo hills), and along the coast: all occur naturally, but most are also associated with slopes engineered by man.

SOIL SLIPS

Soil slips (variously also called popouts, rock-fragment flows, soil avalanches, shallow landslides, and debris slides) are small-scale failures within the weathering and soil mantle (Anderson *et al.* 1959, Corbett & Rice 1966, Bailey & Rice 1969, Campbell 1975). They are probably the most widespread and important type of slope failure in the region (City of Los Angeles 1967). A soil slip usually includes a slip plane, a scarp (a locus for subsequent erosion), a moving slab of material, and/or a semi-fluid tongue of debris on the slope and in the channel (Fig. 1.18). The last-mentioned feature is usually called a debris flow. Slips occur on steep slopes, but not on slopes too steep to retain a debris

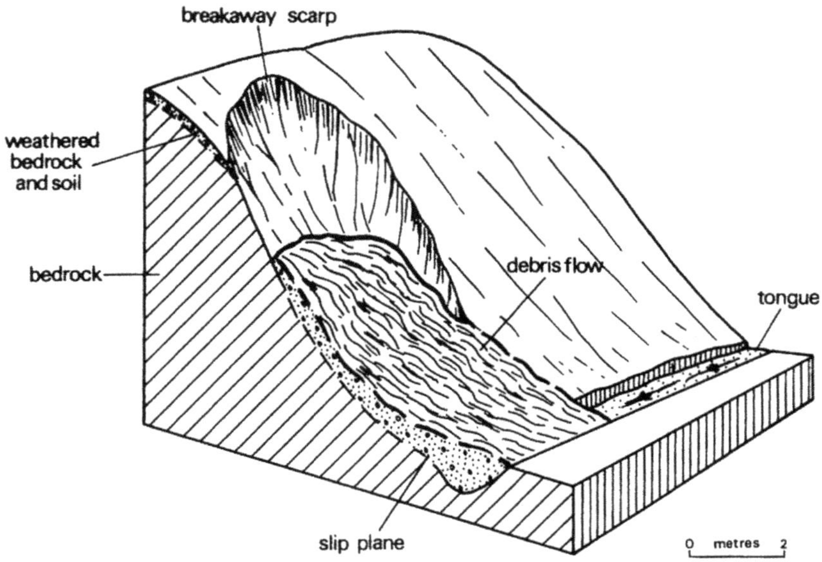

Figure 1.18 The principal features of a soil slip (after Bailey & Rice 1969).

mantle in dry periods or a mantle thick enough to fall in slabs. During the 1969 storms in the Santa Monica Mountains, soil slips occurred on slopes ranging from 40% to 100% (Campbell 1975). Vegetation plays an important rôle, for it both protects and binds the surface, and improves the soil infiltration capacity. Slips are commonest on brush-covered slopes, especially the more sparsely vegetated south-facing ones, and on those recently burned or converted to grass (Bailey & Rice 1969). However, burning does not necessarily increase slippage as it does soil erosion, because of the hydrophobicity which develops in some burned soils. Vegetation also affects the dimensions of soil slips. Campbell (1975) found in the Santa Monica Mountains, for instance, that during the 1969 storms, scars on chaparral-dominated slopes were generally narrower and longer than those on sage, mustard- and grass-covered slopes.

Failure normally occurs where the infiltration capacity of the regolith exceeds that of the rocks below, and when precipitation intensity exceeds the infiltration capacity of the *rocks*. Campbell (1975), in the only thorough study of the mechanism, indicated that in these circumstances a perched water table is produced in the regolith (Fig. 1.19), which will rise, perhaps until the whole surface zone is saturated; seepage flow parallel to the surface may be initiated within the saturated zone; eventually failure will occur when the increased stress locally exceeds the shearing resistance on the regolith.

Campbell (1975, p. 20) described the approximate minimum rainfall conditions for soil slippage as follows:

> an initial period of enough rainfall to bring the full thickness of the soil mantle to field capacity (the moisture content at which, under gravity,

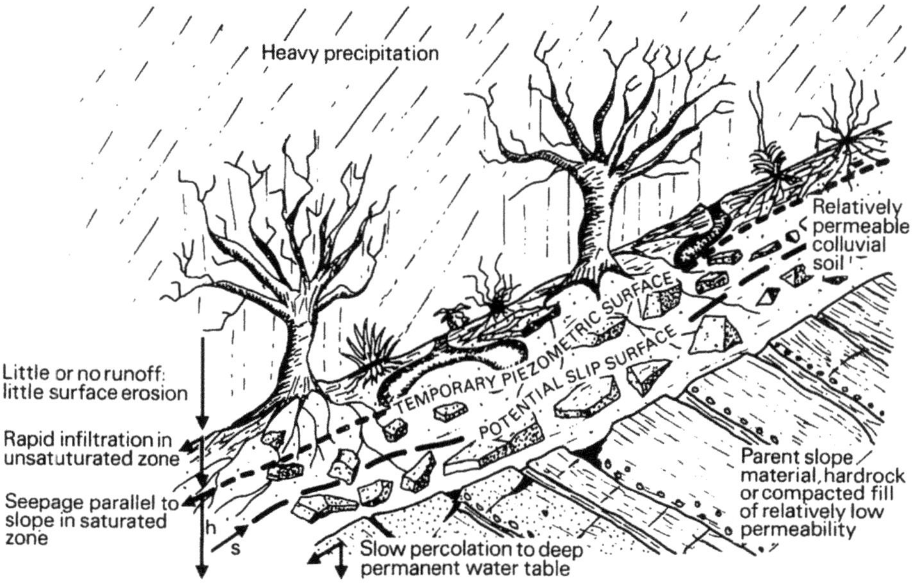

Figure 1.19 The circumstances leading to soil slips in regolith on mountain slopes in southern California. As the piezometric head (*h*) and pore-fluid pressure increase, shearing resistance (*s*) is reduced (after Campbell 1975).

> water will flow out as fast as it flows in), followed by rainfall intense enough to exceed the infiltration rate of the parent material underlying the soil mantle, and lasting long enough to establish a perched ground-water table of sufficient proportional thickness . . . to cause failure.

Evidence from the 1969 storms led Campbell to make the preliminary suggestion that these conditions were reached in the Los Angeles area when the sites of failure had received a minimum total antecedent seasonal precipitation of about 254 mm, after which they were subjected to rainfall intensities of about 6.3 mm/h or more (Fig. 1.20). Storms in the Santa Monica Mountains capable of causing many slips probably have a recurrence interval of about 10–25 years (Campbell 1975), although the frequency with which the antecedent precipitation and rainfall intensity thresholds are crossed and lead to a few failures may be higher, especially at higher altitudes. Slosson and Krohn (1982) similarly suggested that landslides (*sensu lato*) occur when there have been at least five days of high-intensity rainfall, during which at least 177 mm of rainfall are recorded; and that failures were most serious when the five-day, 177-mm phase was followed by the maximum rainfall in a 24-hour period. These empirical thresholds are only general guides, for they certainly vary locally with such specific factors as soil thickness, pre-storm moisture content, vegetation, and slope engineering.

Heavy rains are undoubtedly the most important cause of soil slips, but the

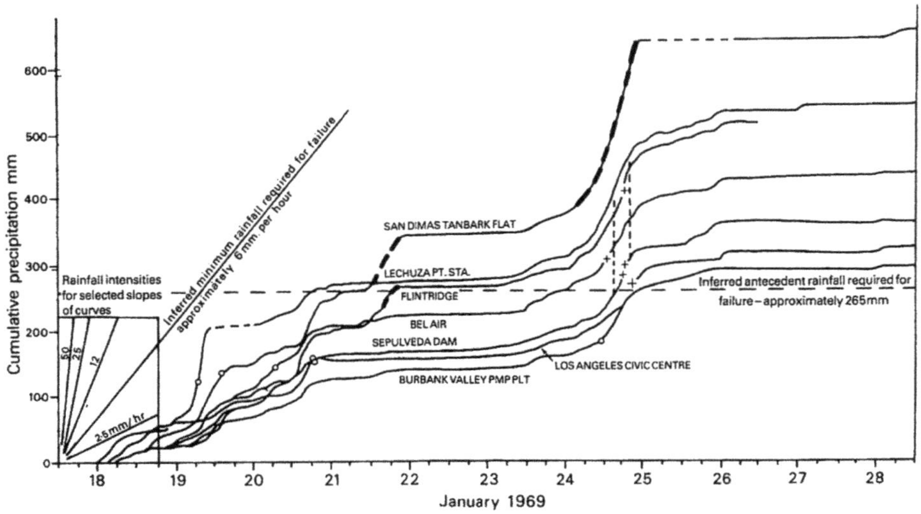

Figure 1.20 Relations of cumulative precipitation, precipitation intensity, antecedent precipitation and timing of slope failures at selected locations in Los Angeles County during the storms of 1969. Thickened dashed lines on the cumulative curves show time range of debris-flow activity (mainly soil slips) in nearby areas; crosses show specific points on curves when debris flows were known to occur nearby; circles show times at which seasonal total precipitation passed 254 mm (after Campbell 1975).

effect of such rains may be enhanced by the removal of chaparral cover, slope oversteepening by channel erosion or human excavation, and by transformation of vegetation cover to grass or bare ground (Brumbaugh *et al.* 1982). Overall rates of debris removal by soil slips are difficult to estimate, but Bailey and Rice (1969) suggested that in the San Gabriel Mountains a rate of 0.177 mm per 11 years (the average return period of the 1965 storm they studied) is a reasonable figure.

Soil slips, and their related debris flows, are a hazard because they are a major cause of fatalities, they damage buildings, related structures and property, they often block drains and other drainage control devices so that drainage is diverted and causes flooding and erosion, and they are major sources of channel debris. And, while the timing of their occurrence may be predictable (and they do tend to concentrate in swales), their specific locations are not.

LANDSLIDES

Landslides usually affect a larger body of material than soil slips, and include both bedrock and regolith. They are mainly a serious hazard because of the destruction they cause to urban fabric on hillsides, especially to lines of communication and property. Unlike soil slips, they are not always triggered by heavy rainfall. If they are, they do not necessarily coincide with the heaviest rainfalls, but arise somewhat later, the time lag being variable. In addition to heavy rainfall, earthquakes provide a trigger for new landslides or the reactivation of old ones. There are two common basic forms in the area: the planar, non-rotational slide, and the rotational slide. Landslides occur throughout the hill and mountain terrain; they are also common along the coast (Pipkin & Ploessel, no date), and in the deeply entrenched alluvial deposits adjacent to it – as in the Pacific Palisades, where some of the most serious urban landslides have occurred.

Fundamentally, landslides come about when the equilibrium between stress on a slope and its strength is disrupted, by an increase in stress, a decrease in strength, or both. Stress is affected primarily by slope height and steepness, the density, structure and moisture conditions of materials, and slope loading. The resistance of the slope, too, is similarly affected by the composition and strength of materials, including its geological structures, the moisture conditions, and the slope's underlying and lateral support. Any balance that exists can be disrupted by increasing stress or reducing strength mainly in the following ways: by increasing the water in the slope (for example, at times of heavy rainfall, by uncontrolled drainage, or over-irrigation), by removing vegetation through urban development or fire (thus influencing soil–water relationships), by slope undercutting or oversteepening (through wave attack, channel erosion, or human slope-grading activities), by tectonic activity, by the imposition of increased loads (for example by placing fill or buildings on a slope), or by reducing the strength of materials through rock weathering.

Landsliding is – and has been through recent geological history (e.g. Stout 1969) – a common phenomenon in Los Angeles County for several basic reasons. Many hillslopes are composed of weak sediments (especially shales, siltstones, mudstones, and otherwise fairly consolidated sediments that include thin clay and silt partings) – Leighton (1966, 1969, 1972) and Pestrong (1976). Swelling clays (like montmorillonite) swell when wet and can initiate instability (Borchardt 1977). Thus, serious slope failures in the mountains are more common in sedimentary-rock areas than in the igneous-rock areas. In addition, there are many potentially unstable, smooth, continuous planes of weakness within and between rock units – bedding planes, faults, unconformities, fractures and foliation surfaces – all of which provide loci for potential movement. Slopes in the mountains are commonly steep and many are potentially unstable. But in southern California many moderate or even gentle slopes (including those in low hills away from the major mountains) are also susceptible to landsliding. In addition, many prehistoric landslides in the area pose a serious threat because they are not easily recognised. Some of them are at present fossil and stable, but they may be reactivated if they are inappropriately developed; others may be motionless at present, but intermittently active. These complex considerations have been incorporated in several attempts to predict landslide location. Figure 1.21 is one such attempt, showing the major zones of landslide potential in the county, and is based on a

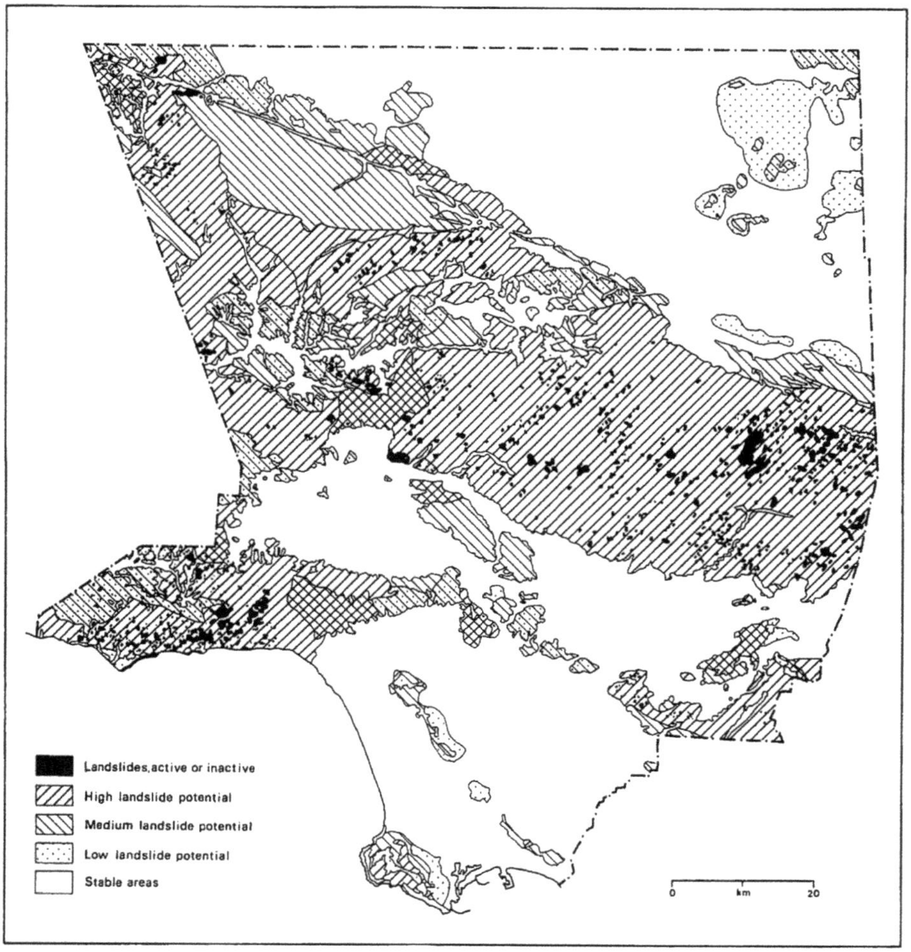

Figure 1.21 Landslide hazard map of Los Angeles County (after Los Angeles County, Regional Planning Commission 1973).

consideration of both historical evidence of failure and the location of unfavourable geological and topographic conditions.

Many landslides occur on natural slopes, but they are also common on slopes engineered by man – cut slopes, fill slopes, cut-and-fill slopes, and temporary construction slopes (Fig. 1.22). The principal causes of man-induced failures (as the work of Leighton (1966 & 1969) and others testifies) are the basal erosion or oversteepening of the toes and/or overloading of the heads of ancient landslides; the creation of instability by cutting new slopes, especially in situations where bedding planes or similar features dip out of the slope; poor drainage or over-irrigation; and the superimposition of unacceptable loads, especially compacted fill and buildings (Fig. 1.22). Of these

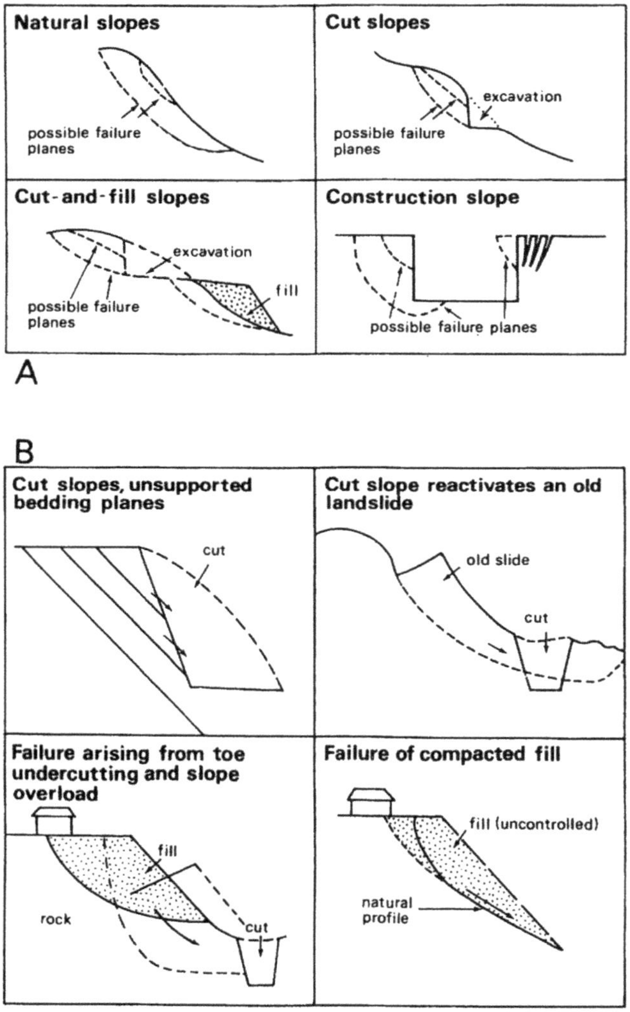

Figure 1.22 (A) Types of slope liable to failure (after National Academy of Sciences 1969). (B) Common slope failures arising from slope engineering in Los Angeles County (based, in part, on Leighton 1969).

causes, the creation of unstable cut slopes is probably the most important; but improper slope and drainage maintenance is also of major significance.

The relative proportions of natural and man-induced landslides in Los Angeles County are not known with certainty, but Leighton's survey suggests that of over 3000 landslides in southern California larger than 30 m in maximum dimension, about 15% were developed by grading activities (Leighton, 1969), and 25–30% were caused by construction activities as a whole (Leighton 1972).

ROCKFALLS

Rockfalls occur where rock falls freely from a slope, usually along bedding planes and joints exposed by oversteepening in material that may be stable in an oversteepened condition even for a short time. They occur naturally in some locations in the mountains, and along the coast (where 'rock' is often Quaternary alluvium); by far the commonest location is in roadcuts in the mountains, especially along fire access trails.

RELATIVE IMPORTANCE OF SLOPE PROCESSES

A key question arises: What is the relative contribution of different slope processes to sediment yield? There have been few attempts to answer it. One suggestion for steep chaparral slopes, proposed by Rice (1973, p. 31), 'with some confidence' but without justification, is as follows:

Process	Contribution (%)
wind	>1
dry erosion	33
soil creep	0
sheet wash }	1
rills and gullies }	11
landslides	54
channel scour	1

These estimates differ substantially from those of the USDAFS (1947) – based on 'the judgement of individuals who were intimately acquainted with the area' – which suggest that in Arroyo Seco 60% of sediment supply came from 'sheet erosion', 25% from 'channel cutting and bank erosion', and 15% from roads and highways. In fact, the question has yet to be satisfactorily dealt with, despite the importance of its answer to conservation and debris-control planning.

Channel flows and sediment yield

The ephemeral removal of channel sediment (mainly during storms) commonly involves a variety of flow types which are primarily determined by the nature and quantity of the available debris, and the ratio of sediment to water. The range is from river flows with relatively low sediment content to viscous debris flows.

River flows need relatively little discussion here because their nature and effects in alluvial channels have been thoroughly studied by geomorphologists and others over many years (e.g. Richards 1982), and their impact on human activities is well understood in most environments. The principal problems of river flow in the major channels of Los Angeles County arise from bank erosion (which can undermine structures adjacent to the channels), scour and fill within channels (which can damage channel facilities, especially bridges), and overbank flow, with its accompanying sedimentation (which can cause inundation of, and damage to urban developments and farmland). As the county has developed during the 20th century, and the river systems have become increasingly controlled, flow in the main channels has become less damaging. But debris flows, and the relations between different types of channel flows, especially on unprotected alluvial fans and in small drainage catchments, pose distinctive problems that require comment.

DEBRIS FLOWS AND ALLUVIAL FANS

Debris tends to accumulate in channels between storm events, perhaps for several years, and when runoff occurs there is often a readily available, abundant debris supply. In addition, sediment is usually supplied to channels from slopes during storms. For example, soil slips and landslides may merge downslope into debris flows as their water content increases and they are remoulded by their own movement. The movement of debris flows (itself highly erosive) may erode channel banks and thereby initiate landslides, thus contributing further to the sediment load of flows. Indeed, such debris supplements, together with temporary debris jams in the channels, have been observed to generate flow surges (Sharp & Nobles 1953, USDAFS 1954, Scott 1971).

Highly charged debris flows occur in storm periods or when snowfall melts, especially in headwater areas where the sediment–water ratio is highest. The flows often appear to be laminar, sometimes with less viscous flows at the surface, but may become turbulent where they are restricted or where the channel surface is rough (Campbell 1975). According to Campbell (1975), it is likely that flows accelerate on slopes over 50%, are of relatively constant velocity on slopes between 50% and 20%, and decelerate on slopes less than 20%. Velocities range from 1 kph to 43 kph.

A mudflow is a species of the genus debris flow. Deposits of both are poorly sorted, often with bimodal size distributions of debris and little or no bedding. But debris flows include less than 10% silt/clay, and are associated with the transport of very coarse material, whereas mudflows contain more than 10% silt/clay. One result of this is that, for the same slope, mudflows move faster and further and are, in general, less damaging; mudflows may also move over very low slopes (Sharp & Nobles 1953). Both debris flows and mudflows leave distinctive evidence: clearly defined fronts, usually of coarse debris, transverse to flow direction (in the case of debris flows, these are often ramparts of large boulders); clearly defined lateral margins (often associated with levées and coarser material); and dissection by subsequent flows with less sediment that can move on lower gradients.

In the mountains, debris flows and mudflows form debris trains, or fill up channels and reservoirs. On the plains, debris flows and channel flows with lower sediment concentrations contribute to the development of alluvial fans.

The fans vary greatly in their size and complexity. For example, numerous small, steeply sloping debris fans are supplied by small, mountain-front catchments (see, for example, Fig. 1.4, p. 6). On the other hand, major catchments and valleys in the mountains (such as Tujunga Wash) leave the mountains in widening valley floors that are similar to alluvial fans on the plains, but which are more confined by the adjacent mountain slopes and are therefore potentially more hazardous; Scott (1973) called these features 'fanhead valleys'.

It is in the fanhead valleys and on the alluvial fans that ephemeral flows create the most serious hazards, for several reasons. First, the nature of an individual flow is largely unpredictable at any one location. Indeed, during a flood, the character of flow often changes from debris flow to relatively clear flow, with unpredictable pulsations from time to time. Second, the type of flow may vary from flood to flood – from perhaps a mudflow on one occasion, to a water-dominated flow on another. Third, the precise route of a flow is, to a degree, unpredictable across complex alluvial fan systems, because flows can be redirected out of established channels by varying sediment loads (especially debris surges), by local deposition that blocks channels (including artificial conduits), and by levée formation (Scott 1971). Even the lower courses of the major Los Angeles and San Gabriel rivers have changed in the last 100 years during floods. Fourth, because flows are rare, they tend to be unexpected and therefore to take communities (if not the management agencies responsible for them) by surprise. Fifth, debris flows can have enormous erosive power, especially on those occasions when their internal dispersive forces carry along huge boulders (Rantz 1970). But the essence of the ephemeral flow hazard relates to the spatial and temporal unpredictability of flow events.

SEDIMENT YIELD

The deposits left by channel flows, especially those that accumulate in debris basins and flood-control reservoirs, constitute the principal measure of sediment yield. Sediment yield (volume per unit area per unit time) is more predictable than the nature of individual events, and has been measured for a number of years. Amongst several attempts to predict regional erosion rates (see Part 3), that of the LACFCD (1959) has been adopted as a basis for minimum debris-basin design capacity (LACFCD 1965). It recognises major sediment yield zones (five in the Los Angeles River Watershed, and four in the Santa Clara Watershed), each of which has a distinctive debris-production curve (Fig. 1.23). The differences between these zones reflect variations in

Figure 1.23 (A) Debris-production zones in Los Angeles County. (B) Relation between debris-production and drainage area for each zone. Zones A–E are in the Los Angeles River Watershed; zones F–I are in the Santa Clara Catchment (after LACFCD 1965).

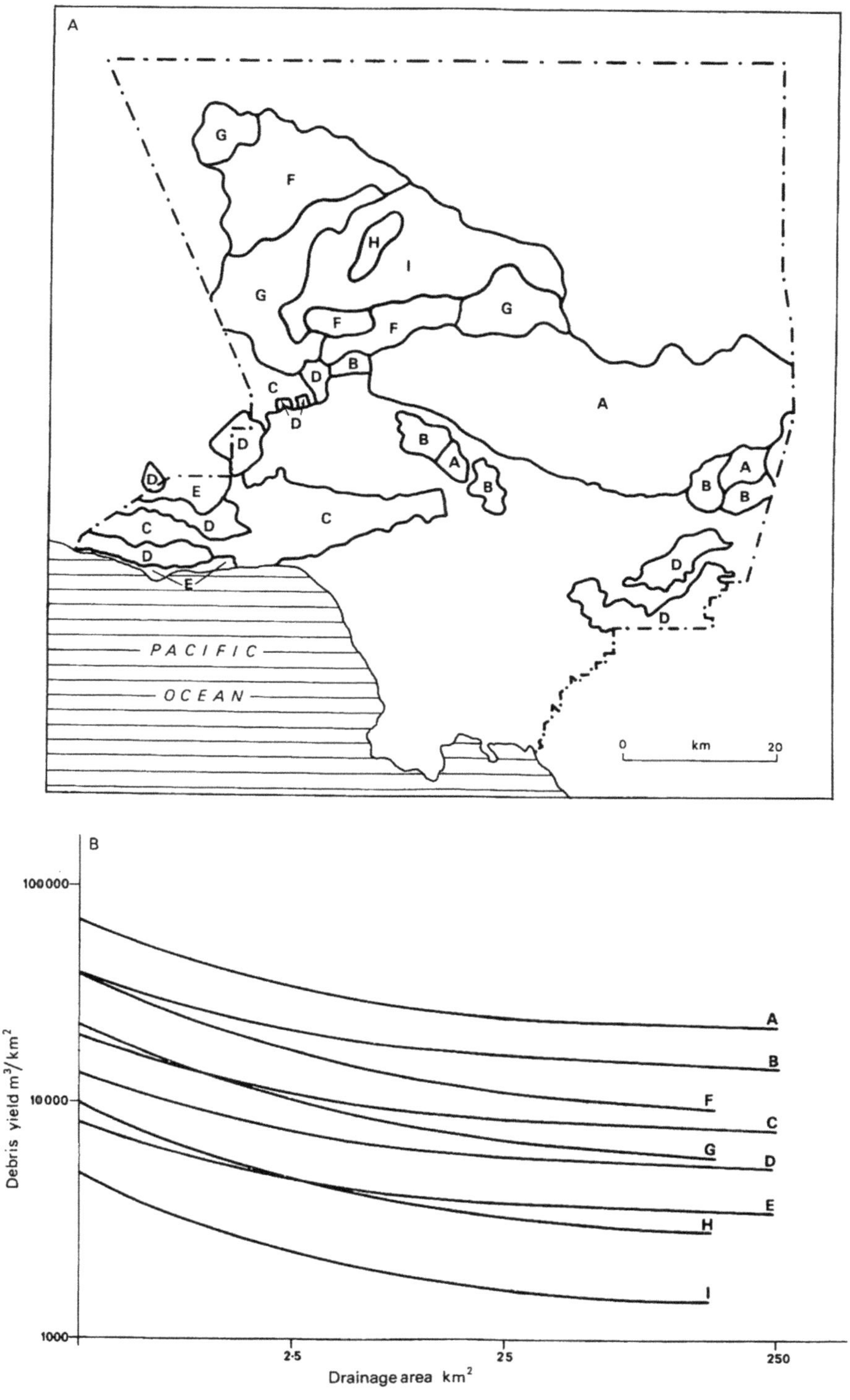
A
PACIFIC
OCEAN
0 km 20
B
100 000
10 000
1000
Debris yield m³/km²
2·5
25
250
Drainage area km²
A
B
F
C
G
D
E
H
I

factors controlling sediment yield, such as geology, aspect, altitude, drainage, vegetation, climate, and fire. In general, sediment yield is inversely proportional to the size of the drainage basin, because storage becomes more efficient, or sediment transfer becomes less efficient, in larger basins (Fig. 1.24, Tatum 1963, Scott & Williams 1978). The highest erosion rates arise from small basins. For example, Scott and Williams (1978) showed that a 1.3-km^2 watershed in an area of high erosion rates has been eroded at an approximate average annual rate of 5900 m^3/yr for the last 40 years (a net rate of 2.3 m per 1000 years). In contrast, the estimated average sedimentation rates in the major flood control reservoirs of larger drainage areas are between 200 and 1947 m^3/km^2 per annum (Kenyon & Coakley 1960). The USDAFS (1974) suggested that in the Los Angeles River Watershed, rates of 737–885 m^3/km^2 per annum are normal.

Short-term events can produce some spectacularly high rates. The highest record of a season's sediment yield in the Los Angeles area is from East Hook Canyon – at least 59 000 m^3/m^2 (Scott 1971). Such high rates are usually accelerated above 'normal rates' by fire and other man-induced changes. The acceleration of erosion rates associated with roads and tracks in the mountains is important in this context. The USDAFS (1974) estimated, for example, that 'road erosion' may contribute in the order of 0.6–15% of total erosion in the Los Angeles River Watershed.

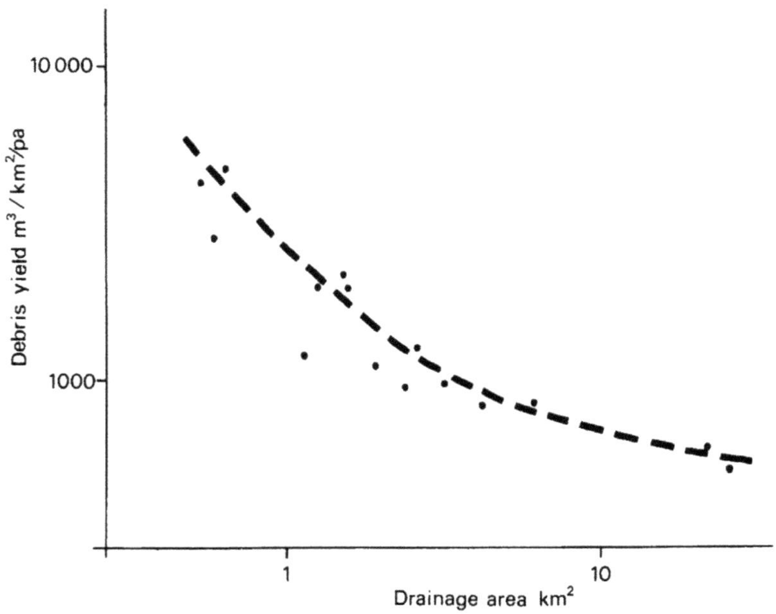

Figure 1.24 Average annual sediment yield rates for a group of catchments without check-dam stabilisation programmes in Los Angeles County for periods of 25–42 years (after Scott & Williams 1978).

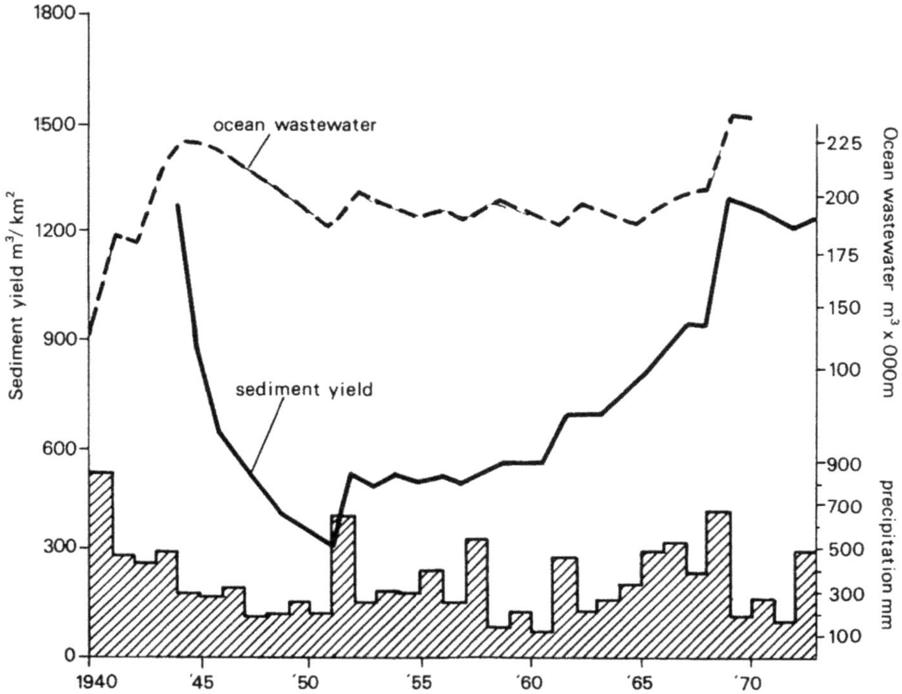

Figure 1.25 Five-year running means of regional annual sediment yield in the Los Angeles River Watershed and total measured waste water to the ocean, and annual precipitation in Los Angeles. (The sediment and waste water data from Ruby 1973; precipitation data from Fig. 1.11).

Another important factor influencing short-term rates is the extent to which debris is stored in the drainage basin ready for removal in a storm. Evidence on this is slight, but the very high yields are probably associated with large sediment stores, mainly in mountain-valley floors. Once such stores are emptied or reduced, the yields might be much less in the next storm, even if it is of the same intensity as its predecessor.

Over a long period, sediment yield will also reflect climatic trends, including dry and wet phases. Ruby (1973) showed, for instance, that there was a *regional* sediment trend in the period 1944–69 which corresponded approximately to *regional* rainfall trends and the quantity of waste water reaching the ocean in the same period (Fig. 1.25).

Most of the sediment-yield data mentioned above refer to sediment collected either in the mountains or at the mountain/plain boundary. Sediment yield is very different at the sea, and it must have changed substantially over time, as the urban area has grown and especially as sediment has increasingly become trapped behind barriers across the rivers. Unfortunately, the supply of sediment to the sea by the San Gabriel and Los Angeles rivers has been only poorly monitored, the main data coming from a few surveys of delta accumulation (Brownlie & Taylor 1981). The available data suggest that the combined average annual sand delivery to the sea from the two rivers may have been approximately 600 000 m^3 under natural conditions, but that today the rate has been reduced to only about 200 000 m^3 (Brownlie & Taylor 1981).

Conclusion: the fundamental problem

In order to make rational decisions, the environmental manager needs to know when, where and why slope failure, slope erosion, runoff, and debris movement will occur. This discussion of the preconditions and environmental changes influencing the geomorphological hazards reveals why it is so difficult to make satisfactory predictions. The natural landscape is dominated by great complexity of form: by innumerable steep, highly active slopes in the mountains, and by many small, relatively independent drainage basins in which the nature of flooding and location of flood routes on alluvial fans is often difficult to predict. And the landscape is underlain by a complex pattern of bedrock, weathering mantles and soils, and numerous relatively weak geological structures – faults, folds, and bedding planes. This complex assemblage of features is cloaked by an altitudinally zoned vegetation that is vulnerable to fires. The fires are started frequently, but in generally unpredictable locations, and their area, timing with respect to storms, vegetal recovery rates, and specific impact on geomorphological activity are uncertain. The broad pattern of land-use changes involving urban development is more predictable – especially in terms of location because of the need for planning permission prior to development – but the broad pattern disguises the local details of interference with geomorphological systems. As a result, the precise location of, say, foundation failure on steep slopes is, within the population of steep and potentially unstable slopes, also largely unpredictable. While engineering works designed to control natural hazards have created a more predictable pattern of change, extreme events may at times distort even this pattern. And possibly most important of all, the principal driving forces of earthquakes and storms are ephemeral and not wholly predictable in terms of their frequency, magnitude, duration, and location.

The difficulty of predicting geomorphological change is increased because this constantly altering environment is attacked by several distinct processes, the relative importance of each of which is uncertain, and which involve different scales of mass transfer, timescales of operation, and distances of movement. Furthermore, each process has its own threshold of activation, and the various thresholds are reached over different time periods. Thus some of them, such as raindrop impact erosion, may be activated in frequent events of low magnitude, whereas others, like slope failure and debris flow, may operate only in high-magnitude, low-frequency storms. In addition, the cumulative effects of minor or continuous geomorphological changes may themselves create a potentially unstable situation in which a threshold may eventually be crossed as a result of even a relatively minor climatic or seismic event. For example, progressive weakening of a rock mass by weathering may lead ultimately to the initiation of a landslide; or the steepening of a surface by slow deposition ultimately may permit gully initiation.

It is the complexity of terrain characteristics, and the spatial and temporal unpredictability of landscape transformations and the forces of geomorphological change that have created, and to some extent still dominate, the management problem. Whilst it may be possible to state statistical probabilities for an event occurring at a particular location based on historical records, the probabilities are in fact constantly changing as the environment is modified, and they often form a relatively insecure basis for prediction. In such circumstances, it is scarcely surprising that progress towards control has been slow, expensive, and politically controversial, and that when storms or earthquakes strike, they can still wreak havoc.

2
CRISES

A SEQUENCE OF STORMS

One of the cardinal features of human response to natural hazard events is that management changes are intimately associated with crises. The degree of response is not necessarily directly correlated with the magnitude of the physical event, but it is correlated closely with the perception of the damage an event causes, especially if human lives or costly litigation are involved. The history of geomorphological hazards in Los Angeles County (as elsewhere) is very much a sequence of crises and responses, in which the crises are the crucible of change and the principal means by which scientific understanding of the phenomena is increased.

The following description of the main geomorphological events in Los Angeles County during the 20th century brings together, for the first time, the data collected by local management agencies and others relevant to each major event. As a community comes to appreciate its problems over time, there tends to be a substantial growth in data collected after each crisis. And data for more recent events tend to be better preserved. Thus the data bases are better for the more recent events. For example, data for the 1914 storms are confined to only a few sources, whereas those for the 1968 storms are so numerous as to be almost overwhelming.

But the data used are probably not comprehensive. The archives often are simply *ad hoc* collections of documents kept in a library or filing cabinet for possible future reference. The material preserved is a selection of that produced originally. How comprehensive are the extant data for each event is unknown, although it is sometimes possible to identify gaps in the archives. Furthermore, it is possible that some important historical documents and archives may have been inadvertently missed in the search. Most of the agencies mentioned in Part 3 have been consulted, and most retain some historical data. Particularly valuable were the libraries of the LACFCD, the USDAFS, the US Army Corps of Engineers (USACE), and various city departments. General reviews (such as those by the USACE) have been heavily used, but, where possible, primary data have also been analysed.

In the following account, each major crisis is described – as far as data permit – in terms of its physical dimensions, the type, location, and cost of damage, and the effect the events had on subsequent institutional responses. The details of the responses are considered systematically in Part 3. The emphasis is placed on major events, although some relatively minor events are mentioned in passing.

DOI: 10.4324/9781003168645-4

1914

In so far as there is a beginning to the modern history of storm-hazard management in Los Angeles County, the storms of 1914 mark that beginning. They were not particularly distinctive or severe, nor were they wholly unexpected. The region had experienced many similar events before. Within living memory, the storms of 1862, 1866 and 1884 were probably larger in terms of rainfall and runoff, those of 1889 were estimated to have been 70% greater, and there had been heavy rainstorms in 1891, 1906 and 1911 (Los Angeles County Board of Supervisors 1915). In his appendix to the County Board of Supervisors (1915) report, C. T. Leeds listed floods in the vicinity of Los Angeles on no less than 41 occasions in the 37 years between 1878 and 1914. Carpenter (1914) recorded floods of one sort or another in the region in 28 of the 37 years since 1877.

But until 1914, the demand for an effective response was muted. It is true there had been some efforts to control flooding. In 1898, for example, a commission was set up to investigate the San Gabriel River, but interest in its findings wilted in the warm, dry years that followed (Los Angeles County Board of Supervisors 1915). The storms of 1911 led the County Board of Supervisors to commission a study by F. H. Olmstead (a well known southwestern engineer), and his timely report of October 1913 was conveniently available for the 1914 storm season and its aftermath. And by 1914 there were seven local flood protection regions, but they were small and ineffective (Los Angeles County Board of Supervisors 1915).

The demand for a response to flooding was muted before 1914 because storm damage since 1889 had been slight. But much had changed between the major storms of 1889 and 1914. The population of the county had increased seven-fold to 700 000. Property had increased in value fourteenfold (Bigger 1959), and it was estimated that the value of the 3885–5180 km^2 of land threatened by flooding was over $500 million in 1914 (Los Angeles County Board of Supervisors 1914). At the same time, drainage facilities had been improved, tending to concentrate flow into established channels and thereby possibly enhancing the potential danger in adjacent areas. Urban development, in the form of relatively impermeable surfaces such as graded roads and buildings, also tended to increase the proportion of runoff from any given storm.

DOI: 10.4324/9781003168645-5

Thus the storms of 1914 attacked a landscape that had been transformed since 1889. They differed from previous storms in another very significant respect. For the first time, there was a concerted political response to public pressure, and this gave rise to the first full survey of damage, to research, to plans for flood control, and to the establishment of regional management. (Some of the original documents have not been found. For example, J. W. Reagan's book *Research, Los Angeles flood control, 1914–1915, vol. 1* seems to have disappeared.)

There were two immediate antecedents to the February storms. First, the fire records (although they are incomplete for the early years of the 20th century) show that the storms were preceded by a fire in at least one area: Haines Canyon was burned in September 1913, and thus became a locus of potentially increased erosion (Olmstead, in Los Angeles County Board of Supervisors 1915). Second, the storms were preceded by 10–12 days of rain in January which effectively soaked the ground, so that when the main phase of heavy storms began after midnight on 18 February, runoff was relatively rapid and substantial.

The four stormy days – 18–21 February – produced a maximum precipitation of 492.8 mm at Mount Wilson (1783 m), and as little as 82.5 mm near the coast at Long Beach. Although precipitation generally increased with altitude (Carpenter 1914), rainfall intensity appears to have been highest on the plains, with the maximum hourly rate of 38 mm being recorded in downtown Los Angeles, and falling at 14.00 hours on 18 February after 13 hours of rain (Grover *et al.* 1917, Los Angeles County Board of Supervisors 1915).

The hydrological consequences of the storms are not well documented because there were few accurate recording stations, and some of them failed. But it is clear that the major rivers and their tributaries carried huge discharges of water and sediment through the system. The peak recorded discharge in the Los Angeles River was 881 m^3/s (4.27 m^3/s per km^2) at Main Street, Los Angeles; and in the San Gabriel River it was 755 m^3/s (8.62 m^3/s per km^2) at Rogers Canyon. And an estimated 2.29×10^6 m^3 to 3.8×10^6 m^3 of sediment were deposited in Los Angeles and Long Beach harbours (Los Angeles County Board of Supervisors 1914).

Although discharges generally increased downstream in the fluvial system, discharges per unit area were much higher in small headwater catchments than in the main channels, especially where antecedent fires had exacerbated the upstream hazard potential. Nowhere was this feature more impressive than in Haines Canyon, where the effects of fire led to augmented runoff of 52.6 m^3/s per km^2 (Olmstead, in Los Angeles County Board of Supervisors 1915).

In 1914 the pattern of inundation and the related pattern of damage were essentially confined to downstream locations and to parts of San Fernando Valley (Fig. 2.1, which is from J. W. Reagan's appendix to Los Angeles County Board of Supervisors 1915). The map in Figure 2.1 shows a broad, simple zone of downstream inundation, and a few locations of flooding

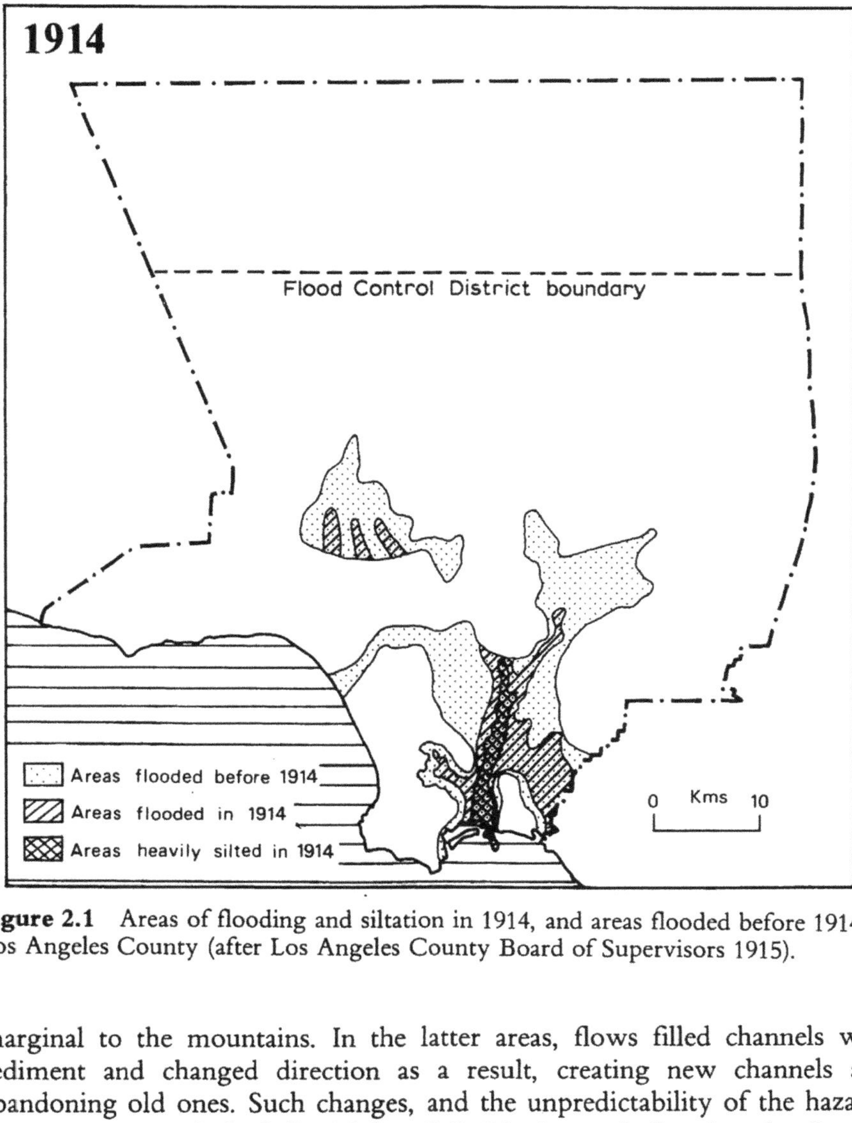

Figure 2.1 Areas of flooding and siltation in 1914, and areas flooded before 1914 in Los Angeles County (after Los Angeles County Board of Supervisors 1915).

marginal to the mountains. In the latter areas, flows filled channels with sediment and changed direction as a result, creating new channels and abandoning old ones. Such changes, and the unpredictability of the hazards they cause, are typical of alluvial fans. J. B. Lippincott (in Los Angeles County Board of Supervisors 1915, p. 103) described this phenomenon near Burbank:

> The area over which the floods wander opposite Burbank is some 7 miles [11.2 km] in width. The maps of the U.S. Geological Survey (Map No. 2) show that in 1893–4 there were five channels in which these waters traveled. The amount of debris brought down from the steeper grades of the mountain canyons is greater than can be transported by the decreasing stream as it spreads over this lane, consequently there is a constant building up the channel to higher elevations, the stream ultimately

breaking through the banks and rushing on to new and unexpected fields, which causes the river continually to shift its channel.

Such changes were not confined to alluvial fans; they also occurred downstream. For example, Reagan's report (in Los Angeles County Board of Supervisors 1914) includes photographs of both new and abandoned channels near Compton.

Reagan estimated (Los Angeles County Board of Supervisors 1914), that damage caused by the floods was at least $10 million, excluding harbour damage (Table 2.1). About a third was attributed to property depreciation and trade disruption, and the remainder was due to losses on agricultural and urban property, state highways, and to losses of private corporations, especially telephone, telegraph, power and railroad companies. The region was cut off from the outside world: there was no telegraph from 18–24 February and rail travel was suspended; Long Beach was an island between the floods of two rivers and the ocean. Very significantly, however, no lives were lost, mainly because main-channel flood waters rose sufficiently slowly for the successful evacuation of people. At least one observer (Carpenter 1914) suggested that the flood losses were more than offset by flood benefits in terms of the replenishing of water supplies. There are no records of slope failures in 1914, but natural soil slips and landslides could well have occurred because the suggested threshold of 254 mm total antecedent precipitation followed by 6.3 mm/h rainfall intensity was crossed.

'Organised flood control was born out of the fears and losses sustained in this catastrophe' (Bigger 1959, p. 2). The first response of the County Board of Supervisors was to appoint a 'Board of Engineers Flood Control' and charge it with the immediate task of reporting within 60 days on the 1914 floods, and with the longer-term task of formulating flood-control plans. The second

Table 2.1 Estimates of flood damages, 1914 (Los Angeles County Board of Supervisors 1915).

I. Physical losses	
A. Ranch and agricultural property in Los Angeles County south of the Old Mission site in San Gabriel Narrows and in San Fernando Valley	$2 626 845
This excludes damage in the San Gabriel Valley and the City of Los Angeles, and is estimated to represent about 60% of the total	
Total estimated damage is therefore	$4 376 000
B. Private corporations (telephone, telegraph, power, railroad companies) and county highways	$2 725 000
C. City of Los Angeles	$500 000
II. Property depreciation, losses due to trade interruption etc., approximately equal to 1/3 of property losses	$2 500 000
Total estimated loss (excluding harbour damage)	$10 000 000

response was to establish a Los Angeles County Flood Control Association which took on the responsibility for preparing legislation. A Flood Control Bill was passed by the State on 12 June 1915, and created the Los Angeles County Flood Control District within the southern part of the county (Fig. 2.1), which combined the responsibilities of the Board and the Association.

The Board of Engineers Flood Control issued two reports (Los Angeles County Board of Supervisors 1914 & 1915) that together reviewed flood damage in general and the events of 1914 in particular, initiated basic surveys (including, for instance, the mapping of the plains at a scale of 1 : 7200 with contours at 5-foot (1.5 m) intervals), and formulated detailed plans for flood control.

The July 1915 plans (summarised in Table 2.2) are significant for several reasons. In the first place, solutions were sought in experience elsewhere, primarily in superficially similar European environments such as the Alps. This was a weakness that subsequent flood experience had to rectify. A second feature of the plans was their concern to protect existing rather than future developments, and to do so by adopting a wide range of fairly modest proposals. They included a scheme of check-dam construction in headwater channels – an idea apparently promoted by Olmstead's experiment in the Western Empire Canyon branch of Haines Canyon, in which check dams were shown to reduce the velocity and quantity of runoff. 'Rebrushing' and 'reforestation', together with dam construction, were also recommended in the headwater areas. Down-

Table 2.2 Summary of flood-control proposals, 1915 (Los Angeles County Board of Supervisors 1915).

Area, type	Cost ($)	Cost of rights of way purchase ($)
Mountain district		
(a) Impeding dams, contour furrows, water spreading	2 381 300	634 500
(b) Channels	817 400	35 700
Total	3 198 700	670 200
San Fernando Valley		
Channel management and water spreading	1 508 400	585 600
Santa Clara drainage		
Channel management	1 619 500	601 900
San Gabriel district		
Check dams, water spreading, levées and other channel improvements	4 381 000	583 000
Coastal Plain district		
Channel modifications and diversion of Los Angeles River	3 625 200	1 829 400
Total		16 508 900

stream, the emphasis was on water-spreading grounds, channel improvement (levées etc.), new and more efficient channels and drains, and redirection of flows in the lower Los Angeles and San Gabriel rivers. The plan thus included a recognition of all the main solutions to the flooding and sediment problems; since that time, the essential changes have been to priorities, the relative importance of solutions, and the design, scale, and location of engineering works.

A third feature of the plans lies in the fact that they contain the seeds of the 'upstream–downstream' controversy: whether it is better to manage the discharge of water and sediment in the headwaters or on the plains. The controversy does not require a simple one-or-the-other solution, because both approaches have some merit, but it does involve conflicting environmental judgements and even rivalry between different management agencies. J. W. Reagan began to criticise the July 1915 plans, with their upstream–downstream balance, in letters and in his appendix to the report. His dissent culminated in his appointment as chief of the LACFCD, against strong opposition, in 1915. Reagan was particularly worried about the effectiveness of upstream structures, and in general favoured downstream protection. For instance, he asked (Reagan 1915):

> . . . is it not probable that reservoirs would be entirely full before the peaks of the floods came?
> and
> If the dams and reservoirs are intended only for obstruction rather than storage purposes, will they not at once fill up with rock and debris to their full capacity?

He was also worried about dam safety and the fate of sediment in areas where gradients and flow velocities were to be reduced. C. T. Leeds pointed out that (as the rainfall intensity data suggest) runoff may be initiated by 'downstream' rainfall, and that water spreading (by saturating the ground surface) might in certain circumstances actually promote runoff. Reagan and others (perhaps with an eye to political support) argued that there was a compelling case for downstream protection of the fundamentally important harbours and of the estimated 532 000 ha of land (see Fig. 2.1) that had been flooded in the past or could be inundated in the future.

Once Reagan was in control, he formulated new plans (Reagan 1917) which – while they retained the same four elements of dam construction in mountain areas, bank protection and water spreading, straightening and modification of major channels, and protection of harbours – shifted the emphasis from upstream to downstream works, and changed the proposed route of the Los Angeles River diversion project.

A final feature of the plans is the recognition that the 'urban effect' was going to make the flooding problem worse. C. T. Leeds summarised the argument as follows (in Los Angeles County Board of Supervisors 1915, p. 174):

> There is still another factor in the destructiveness of a flood, which has not been mentioned but which is of extreme importance and hence must not be overlooked. By the growth of cities and towns, with their great areas of roofs and paved streets, by the extension of paved highways and improved watercourses, and by the return water from increasing irrigation . . . the runoff resulting from any given rainfall is steadily increasing, and will change moderate floods of the past into serious floods in the future.
>
> It must also be foreseen that in the future, uncontrolled runoff will do far more damage through erosion and silting than has been done in the past. For whereas in the earlier history of this county flood waters could spread harmlessly over wide areas whose soil was bound down by grass and other vegetation, these waters now are confined by boulevards, embankments and other restrictions provoking scour.

Reagan (1917) emphasised the first point graphically, showing that when the first graded road is placed round a section, 1.88% of the area is hardened and made relatively impermeable; when the section is quartered, 3.8% is affected; when 10-acre (24.7 ha) sections are created, 14.5% is in such surfaces; and when it is in 2.5-acre (6.17 ha) plots, 28.3% of the original surface area becomes streets and roads.

The plans did not wither in the sunshine following the storms. More floods and $4 million of damage in 1916 revived memories and ensured action. Despite intense political controversy (Bigger 1959), bond issues were approved in 1917. Most of the plans had been implemented by 1924, but they were extended and further finance came mainly from a 1924 bond issue.

1934

The year of 1933 ended, and 1934 began memorably in Los Angeles County. The memorable event – a classic example of alluvial fan flooding in La Cañada Valley – was not a regional crisis (indeed it was highly localised) but it acquired a significance well beyond its area of damage. There are relatively few scientific accounts of it, but that by Troxell and Peterson (1937) is outstanding. Unless otherwise stated, most of the details in the following description are drawn from their study; other major sources include the work of Eaton (1936), Kraebel (1934), Taylor (1934), and Chawner (1935). There are some inconsistencies in the data in these papers, but they are not serious.

La Cañada Valley – extending from Tujunga in the west to Arroyo Seco in the east (Fig. 2.2) – lies between the San Gabriel Mountains, and the Verdugo and San Rafael hills. It is dominated by a broad alluvial plain sloping away from the San Gabriel Mountains which comprises a complex system of entrenched alluvial fans derived from several small mountain catchments. These fans ultimately drain into Verdugo Wash, a generally dry tributary of the Los Angeles River which lies close to the base of the Verdugo Hills.

In 1934, drainage systems of this area dramatically demonstrated their hazardousness. There are 16 small mountain catchments in the San Gabriel Mountains. They have a relief of some 600 m in a horizontal distance of less than 7 km, and collectively they comprise only about 25 km^2. They feed alluvial fans through their entrenched feeder channels, which are up to 50 m deep. The limit of entrenchment is marked approximately by Foothill Boulevard. Beyond, the debris cones are dominated by ill-defined, irregularly used distributary channels. The gentle slopes were seen as ideal for citrus groves, vineyards, and settlements such as Montrose, La Cañada, La Crescenta and Tujunga. As a result, a neat grid of human activity had been superimposed on the area – an apparent triumph of geometry over geography.

The floods had two important immediate antecedents. First, nearly all the mountain catchment slopes had been denuded of vegetation by fire in a four-day period during November. The 'Pickens Canyon fire' covered nearly 2000 ha. More generally, if it is assumed that vegetation takes ten years to recover from a fire, Eaton's (1936) data lead to the observation that a considerably greater area was still suffering from fire effects. Second, the years preceding the fire had not experienced heavy storms, only minor ones. Thus

DOI: 10.4324/9781003168645-6

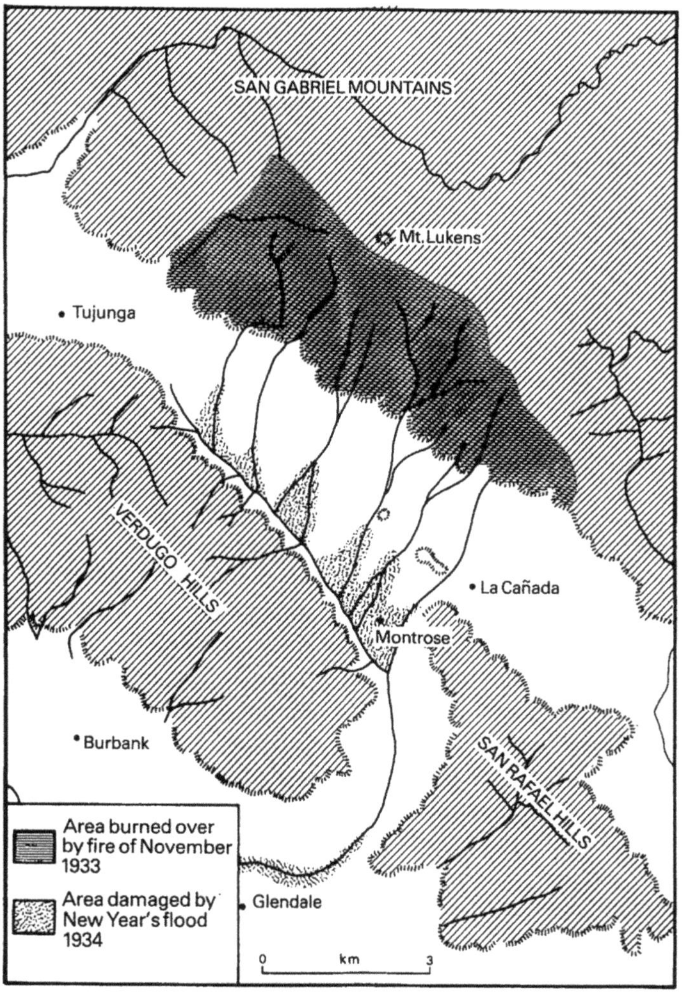

Figure 2.2 La Cañada Valley, showing areas burned in 1933 and damaged by flooding in 1934 (after Troxell & Peterson 1937).

there had been a long period for debris accumulation in canyon floors, a process assisted by the construction of check dams.

The winter storms occurred in two main phases. There was an early storm on 14–15 December, in which 81.2 mm of rain fell. This storm helped to saturate the ground surface, and perhaps to move sediment into canyon channels.

Then, according to Kraebel (1934), 319 mm of rain fell in the Verdugo area in about 38 hours from the early morning of 30 December to the afternoon of 1 January. The major storm within this second phase began on the afternoon of 30 December and continued intermittently for about 18 hours (Troxell &

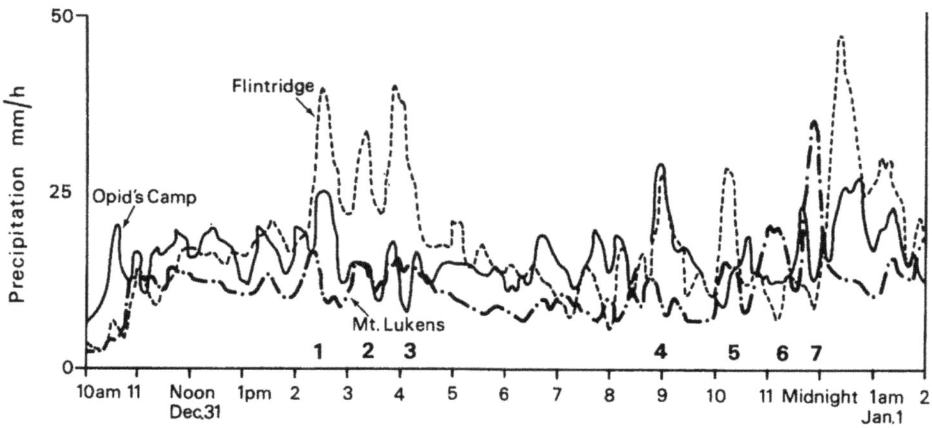

Figure 2.3 Precipitation intensity (based on progressive mean rates for each three five-minute periods from 10.00 h on 31 December to 02.00 h on 1 January) at selected stations during the storms of 1934. Heavy numbers **1–7** identify the major intense showers within the storm period (after Troxell & Peterson 1937).

Peterson 1937). The storm was one of the heaviest on record, increasing the highest 24-hour precipitation values at many stations in the Los Angeles Basin. For example, the station in downtown Los Angeles (with 56 years of observations) had its previous record of 130 mm (1913) raised to 187.2 mm (Daingerfield 1934). But the storm was outstanding in another respect. While rainfall intensities were not particularly high and were, as usual, variable in both space and time, it is clear that rainfall was heaviest in the *lower* foothills of the San Gabriel and Santa Monica mountains. For example, Hogee's Camp (altitude 807 m) received 505 mm; and Topanga Canyon (altitude 227 m) 407 mm, whereas Mt Wilson (at 1783 m) received 396 mm. Rainfall was relatively very heavy in the foothill area tributary to La Cañada Valley (Daingerfield 1934). Rainfall intensity varied in the La Cañada region in both time and altitude (Fig. 2.3). It exceeded a rate of 25.4 mm/h on no less than six occasions at Flintridge (a site above Glendale at 402 m). Troxell and Peterson (1937) recognised seven very intense showers during the storm. Whereas the earlier showers were more intense at lower elevations, the last two – which caused the *dénouement* of the crisis – were actually more intense at higher altitudes.

Conditions and events in the mountain catchments had no eye-witnesses. But a careful reconstruction by Troxell and Peterson reveals that in all probability sediment was progressively moved in pulses down the canyons by the succession of peak rainfall–runoff events on 30–31 December, so that by nearly midnight on 31 December, much debris had accumulated near to the canyon mouths. It was then that these masses of material were struck by the last and most intense rainfall (an intensity of 54.6 mm/h was recorded at 23.47 h at Flintridge), and the poorly sorted mixture of coarse and fine material (including boulders weighing many tonnes) moved *en masse* down the entrenched channels as debris flows. These flows cleaned out sediment in many of the channels, lowering them by over 2 m in places.

Some of the sediment had certainly accumulated during previous years, but some of it may have been moved into the channels during the storms of 1933–34. The question arises as to how much of the debris was generated by erosion and slope failure *during* the storms. The answer is perhaps surprising. Troxell and Peterson noted that to cause wet erosion, an intensity of possibly 25 mm/h may be required. This intensity was exceeded several times between 30 December and 1 January (Fig. 2.3), and this fact, together with photographic evidence of fresh gullying, suggests that slope erosion may have contributed to the debris in the flows. But the intensity was only exceeded at higher elevations in the later showers of the storm, so the contribution may have been little and late.

Soil slips, which undoubtedly also occurred (see Troxell & Peterson 1937, Plate 34B), may also have been too late to contribute. Campbell (1975) suggested that an approximate threshold for soil slips arises when a site has received 254 mm of rain followed by rainfall intensities of about 6.3 mm/h. This threshold was only reached in the La Cañada area at about midnight on 31 December/1 January, and was sustained for about two hours. Thus it could be that the 1934 soil slips provided sediment only for the *next* floods. However, it is possible that the threshold may have been lower and reached earlier, given the mid-December rains and the antecedent fire. If larger slope failures occurred or were reactivated, their locations and timing of movement are unknown.

Check dams – so favoured as a conservation measure by Olmstead in 1915 – were not very successful in retaining debris in 1933–34. While the elementary wire and rock mattress structures seem to have stored sediment effectively during small flows, when the big flood came many of them were destroyed (partly, perhaps, by large boulders rafted on debris flows). Once destroyed, they actually added sediment to the flows: they made the problem worse.

There is strong evidence that the debris flows from the canyons occurred when another critical threshold was crossed. Up to about 21.00 h on 31 December flows were relatively free of debris, and little damage was done. Later, however, eye-witnesses recorded walls of water and debris, perhaps over 7 m high in places, advancing down canyons and the entrenched feeder channels, and moving beyond them between 23.56 h and 00.15 h on 31 December/1 January. Once beyond the entrenched channels, the flows spread out. They lasted for up to 20 minutes. It seems likely that swift streams of clear water, derived from the exceptionally heavy last shower, moved into and over the debris accumulations at canyon mouths, rapidly mobilising them. The debris flows moving *en masse* may not have been very erosive (as evidenced by houses buried, but otherwise undamaged) although boulders weighing 70 tonnes were transported. The flows clearly developed very steep transverse slopes when rounding bends, for in places banks were overtopped on one side of the channel, but the debris levels on the other side were as much as 6 m lower, and well within the original channels. However, following behind these flows, water was underloaded (because little sediment remained to be moved) and highly erosive. Such flows caused very substantial damage to vehicles, property and roads. The event seems to have included a series of 15 or so pulses, and flow velocities varied during the event from about 1.5 m/s to 3.0 m/s (Eaton 1936).

Discharge can only be calculated approximately, because gauging stations in the area were destroyed. Discharge from Pickens Canyon may have reached 7.05 m^3/s per km^2 (Troxell & Peterson 1937), and maximum sediment yield ranged from 13 279 m^3/km^2 in Pickens Canyon to 19 771 m^3/km^2 in Haines Canyon (Kraebel 1934, Troxell & Peterson 1937). In total, some 410 000 m^3 of debris spread out below Foothill Boulevard to cover or erode about 20% of the La Crescenta–Montrose area (Eaton 1936). Thus a 7.7-km^2 area was covered by material drawn from only about 25 km^2 of catchment area. Chawner (1935) estimated the total volume of debris to be 534 800 m^3, and to represent watershed erosion of 6.35 cm.

The localised flood – comprising first clear flow, then debris flow, followed by highly erosive water flow – killed over 40 people (39 confirmed; 45 missing), destroyed 198 houses and rendered uninhabitable 203 others, and damaged streets, bridges, highways, vehicles and other fabric of an urban community. Of many instances of the effect of the flows, the Montrose Legion Hall provides a good example: the debris flow crashed through the middle of the building, leaving holes in the uphill and downhill walls, and carrying 12 people, seeking shelter, to their deaths (US Department of Agriculture, USDA 1938). The estimated cost of damage to property was $5 million, to which must be added the cost of $425 000 for debris removal (Pickett, in Eaton 1936), and as much as $4 233 821 for subsequent engineering works designed to protect the La Crescenta and Montrose area (LACFCD 1935).

The floods of 1934 were not confined to La Cañada Valley. There was high flow in many valleys, with reports of damage in, for example, Topanga Canyon (Santa Monica Mountains). Ballona Creek, and Brand Canyon on the western side of the Verdugo Hills. In the last-mentioned, 17 out of 19 check dams were destroyed, and 61 120 m^3 of debris spread over Glendale (LACFCD 1958), but it was only in the La Cañada area that damage was serious – the area where mountain catchments had been burned in the previous autumn.

The 'New Year's Day Flood' (as it came to be known) profoundly affected local perception of environmental hazards for several reasons. Most dramatically, the flood struck as the bells tolled in the New Year, and the slaughter of people, many of them revellers, created banner headlines on a day traditionally given to optimism about the future. Its impact was further enhanced because it came without warning, in the dark. And it ended the relatively long dry period that followed the harrowing events of 1914 and 1916. While there had been minor floods since 1916 – in the fire-denuded Fish

Creek in 1924 and in Sunset and Brand canyons in 1928, for instance – it was the first important test of control measures introduced mainly by the LACFCD following its establishment in 1915. It showed them to be woefully inadequate, and to be seriously lagging behind urban development. Another reason for its impact was that it came after a period of rapid agricultural and urban development that had done much to create the warm, dry, sun-ripened image of the region, an image unblemished by storms. Since 1914, property values in the county had risen several hundred per cent to over $4000 million. Population had increased to over 2.25 million of whom it is estimated only 8% had experienced, or had an appreciation of, the 1914 flood (Eaton 1936). By 1931, land liable to inundation included property valued at over $300 million, and 380 000 people (Eaton 1936).

The flood was, coincidentally, a timely warning, for it came just before the period when federal authorities were seeking ways of constructively alleviating unemployment, and when the federal government was under pressure to become further involved in flood management – pressure that culminated in the *Flood Control Act* of 1936. This act was to have as profound an effect on the county as that which created the LACFCD. The La Cañada disaster is used to support arguments for flood control in the Los Angeles region in almost all subsequent public documents. It was not only a stick with which to beat more funds from government and populace, it was also a scar on folk memory comparable to that of the Long Beach–Compton earthquake in the preceding year (e.g. Oakeshott 1973).

The La Cañada flood was important in one other respect. It had all the characteristics of a severe alluvial fan flood in southern California: it followed a period of debris accumulation in channels, fire in the foothills, and heavy antecedent precipitation; it combined both spatial and temporal variability in the movement of water and sediment, and it surprised innocent citizens of the balmy plains below the entrenched feeder channels. It was an event that should have been predicted. Indeed, it was, by J. A. Bell and H. Hawgood in 1925, in a consultants' report to Los Angeles County Board of Supervisors, 31 December 1925 (quoted in Troxell & Peterson 1937, p. 69):

> La Cañada valley exhibits to an unusual degree the effects of violent flood action. The whole area, ringed about by steep mountains, is boulder-strewn virtually down to the main drainage channel, the Verdugo Wash. The character of this boulder-strewn alluvium is significant. Its great size, at least half a mile down the slope from the mouths of the canyons, is clearly indicative of but one thing – periodic flood discharges of extremely violent nature.

1938

If the floods of 1934 were a localised warning, the storms of 1938 both justified previous actions and reinforced arguments for further action. Their timing was perfect, coming shortly after the initiation of programmes under the 1936 *Flood Control Act*, at a time of reappraisal. They were also a large, regional phenomenon, as big or bigger than the storms of 1914, and they were comparable to the computed capital flood peak based on a 50-year rainfall frequency: the whole of southern California was affected, not simply a small area of Los Angeles County. Furthermore, the events of 1938 were thoroughly chronicled, most notably by the United States Geological Survey (USGS; Troxell *et al.* 1942), the USACE (1938), the LACFCD (1938), the Conservation Association of Los Angeles County (1938), and the USDAFS (1939). The following description is based on data in these and related reports.

The rural and urban development of the county had not been dramatic since 1934, given a generally adverse economic climate. The population had grown by only 200 000, to 2 450 000, since the La Cañada flood. However, the physical antecedents to the floods were more significant. First, the preceding years had been relatively dry (the 1934 storms not withstanding), so that debris accumulation had been largely confined to the mountains, with relatively little clearout on to the plains. For example, the San Dimas Creek flood-water control reservoir lost 22.8% of its capacity through sedimentation between its construction in 1922 and 1938, but approximately 18.1% of this reduction occurred in the 1938 storms (Troxell *et al.* 1942). Additional debris storage had been provided in the years preceding 1938, especially in the form of debris basins such as those in Dunsmuir, Shields and Pickens canyons which were constructed following the 1934 catastrophe in the La Cañada area.

Unusually, antecedent fires do not seem to have been an important factor. Between 1934 and 1937, the records reveal few fires (Table 2.3), and of these only those near Mount Lowe would have significantly influenced sediment yield in the major river catchments (as they probably did in Arroyo Seco, see below); the Point Dume fire might have affected erosion on the lower slopes of the Santa Monica Mountains.

Precipitation prior to the main storms was, as is so often the case, a particularly important factor, but it took an unusual form. Los Angeles received only 0.76 mm of rain in the 192 days up to 8 December. A moderate

DOI: 10.4324/9781003168645-7

Table 2.3 Fires, 1934–37 (from original records held by the LACFCD).

Date	Area (ha)	Approximate location
1934	—	Brown Mountain (Mt Lowe–Arroyo Seco area)
1935	9625	Point Dume (Santa Monica Mts)
	781	Mt Lowe ('Altadena fire')
1936	808	Mt Lowe area (San Gabriel Mts)
	549	San Francisquito (N of County)
1937	357	Mt Baden Powell (NE San Gabriel Mts)
	751	Bouquest Reservoir (N of County)

storm on 9–12 December broke the drought, and was followed from 1 to 25 February by precipitation that was greatly above normal and more than served to restore the seasonal deficiency: this wet spell undoubtedly created a well-saturated soil prior to the main storms at the end of the month. Some of the precipitation fell as snow in the mountains above 1500 m, giving an average depth of some 457 mm before the main storms began (J. J. Prendergast in Troxell *et al.* 1942).

The main storms occurred between 27 February and 3 March. Troxell *et al.* (1942) analysed their characteristics in great detail and distinguished four major phases:

(a) 27 February to early 28 February: light, intermittent rain, and continuous intense precipitation for most of the day on the 28th, with very high, short-duration intensities, especially near the coast;
(b) 1 March: no appreciable rainfall;
(c) 2 March: extremely heavy precipitation with the maximum 24-hour rainfall being the greatest on record at most mountain locations;
(d) on 3 March: light, intermittent rains until early afternoon.

In short, it was a storm period of two major rainy spells.

Rainfall amounts and intensities varied temporally and spatially; in particular, the loci of the storms moved eastwards during the period and, unlike the storms of 1934, precipitation generally increased in intensity and amount with altitude, the front ranges of the San Gabriel Mountains taking the brunt. The highest total rainfalls were recorded in the San Gabriel Mountains at Kelly's Camp (altitude 2529 m; 817.9 mm total precipitation), and at Hogee's Camp (altitude 838 m; 782.3 mm total precipitation). Most areas of the county received over 570 mm in the season, although it is clear from the analysis by Troxell *et al.* (1942) that there were considerable, topographically controlled variations in precipitation in the mountains. Rainfall intensities were greatest (Fig. 2.4) late on 28 February and in the daytime of 2 March. The periods during the storms for which precipitation exceeded some assumed

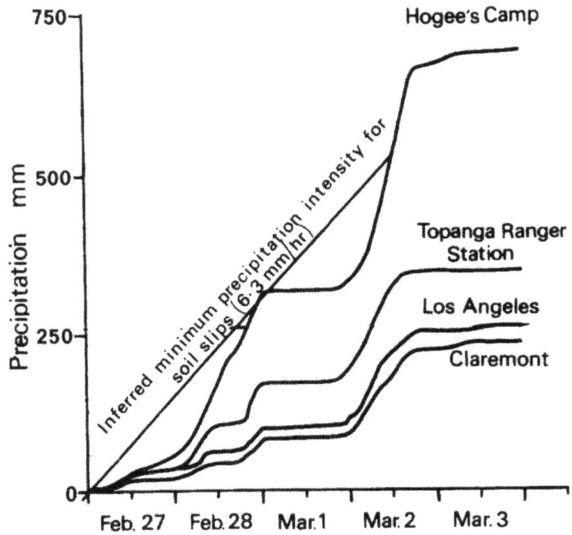

Figure 2.4 Cumulative curves of precipitation for the storms of 1938 at recording stations representative of the mountains (Hogee's Camp), the foothills (Claremont), the plains (Los Angeles), and the coast (Topanga) (after Troxell *et al.* 1942).

geomorphological thresholds are shown in Table 2.4. In general, precipitation rates for 30-minute periods or less were lower than previous records. For example, the maximum 5-minute rate at Los Angeles in 1938 was 63.5 mm/h, a figure exceeded 17 times previously during the period of record. The storms were distinguished, therefore, by their antecedent rainfall, longevity, two-phase character, sustained spells of heavy rain, and intensity in the mountains.

Evidence of slope damage in the mountains, by comparison with channel and floodplain damage on the plains, is not well chronicled. The Los Angeles County, Department of Forester and Fire Warden (1938) estimated damage to fire protection facilities alone in Forest Service and County forest lands to be $914 662. They recorded hundreds of slope failures (especially on the south slopes of Mount Wilson), and extensive erosion and deposition associated with 1662 km of truck trails, 1052 km of fire prevention telephone lines, 1303 km of pack trails, 1408 km of fire breaks, and various administration sites and water-storage locations. It seems that 'not one truck trail or pack trail of the Forest Service and County system was left in usable condition' (p. 10).

The only available systematic record of such slope damage was undertaken by the USDAFS (1939), who surveyed the 77 km^2 of Arroyo Seco above Devil's Gate Dam. The detailed survey showed that the primary causes of huge sediment loads were fourfold: channel erosion and gully scour (unquantified); landslides (including soil slips), which produced an estimated 147 420 m^3 of debris (Fig. 2.5); and 'road erosion' (i.e. erosion from road cuts

Table 2.4 The storm of 27 February to 4 March 1938: selected precipitation data and geomorphological thresholds.

Station	Altitude (m)	Total precipitation (mm)	Hours with precipitation intensity above 7.6 mm/h[a]	Total precipitation (mm) at intensities above 7.6 mm/h	Maximum precipitation intensity (mm/h) per 5-minute period
Hogee's Camp (San Gabriel Mts)	838	754.6	36	347	91.4
Mt Wilson (San Gabriel Mts)	1783	666.0	28	266	76.2
Pickens Canyon (San Gabriel Mts)	1295	440.0	21	129	60.9
Rossmoyne (Verdugo Hills)	457	330.0	13	83	121.9
Topanga Canyon (Santa Monica Mts)	227	378.0	15	118	73.1

[a]The value nearest to 0.25 inches in the available data, assumed to be related to the threshold of soil slippage.

Primary data: Troxell *et al.* (1942), based on rain-gauge measurements between 27 February and 4 March.

and unpaved road surfaces), which generated an estimated 226 800 m^3. The last two categories produced an estimated sediment yield of 4818 m^3/km^2. No doubt this damage was fairly typical of the mountain area as a whole, although yields in Arroyo Seco may have been somewhat higher than elsewhere as a result of antecedent fires (see Table 2.3).

That the conditions for soil slips and erosion were achieved is beyond doubt. For example, antecedent precipitation of 254 mm was achieved in places before or soon after the beginning of the main storm period, and rainfall intensities of 6.3 mm/h were attained for many hours (Table 2.4). As these thresholds were crossed in the first phase of the storm, it is probable that much sediment was provided to the channels to supplement earlier accumulations *before* the major efflux of 2 March.

Damage in Arroyo Seco was estimated at $437 000 (comprising $170 000 road damage; $171 150 reservoir sedimentation; $28 000 damage to drainage diversion works; and $68 200 to property). The Arroyo Seco Dam itself was overtopped in 1938, causing $1 335 000 of damage downstream, especially to roads and bridges ($497 000), and utilities ($444 000), and less significantly to business and residential property ($85 000) and agriculture ($7000).

Channel discharge is very largely associated with the two main precipitation peaks, but nearly all the flood runoff, and certainly peak flood runoff, was associated with the second rainfall peak on 2 March. For example, flood control reservoir records in Los Angeles County showed that during the maximum 24-hour precipitation period, 66% of the entire flow for the five-day period occurred. Despite problems of monitoring discharge during flood conditions, Troxell *et al.* (1942) accumulated an impressive record of discharge based on stream-gauging and reservoir inflow–outflow data. They showed that the four rainfall phases were matched by similar runoff phases. But the lag times between rainfall and runoff peaks varied greatly, according to such factors as drainage basin geometry (slope, area etc.) and distance downstream. For example, lag times for the West Fork of the San Gabriel River were between 2 and 3 hours for periods of 6 hours or less; for longer periods than the maximum 6 hours, the lag increases to about 5 hours for the maximum 24-hour flow period over an estimated flow distance of approximately 8.0 km; lags increase, of course, down valley, and flood peaks are both reduced and delayed by reservoir storage. Nevertheless, warning times were likely to have been relatively short, and would have contributed to the impact of the flood.

Rates of runoff were exceptionally high in large catchments, often exceeding

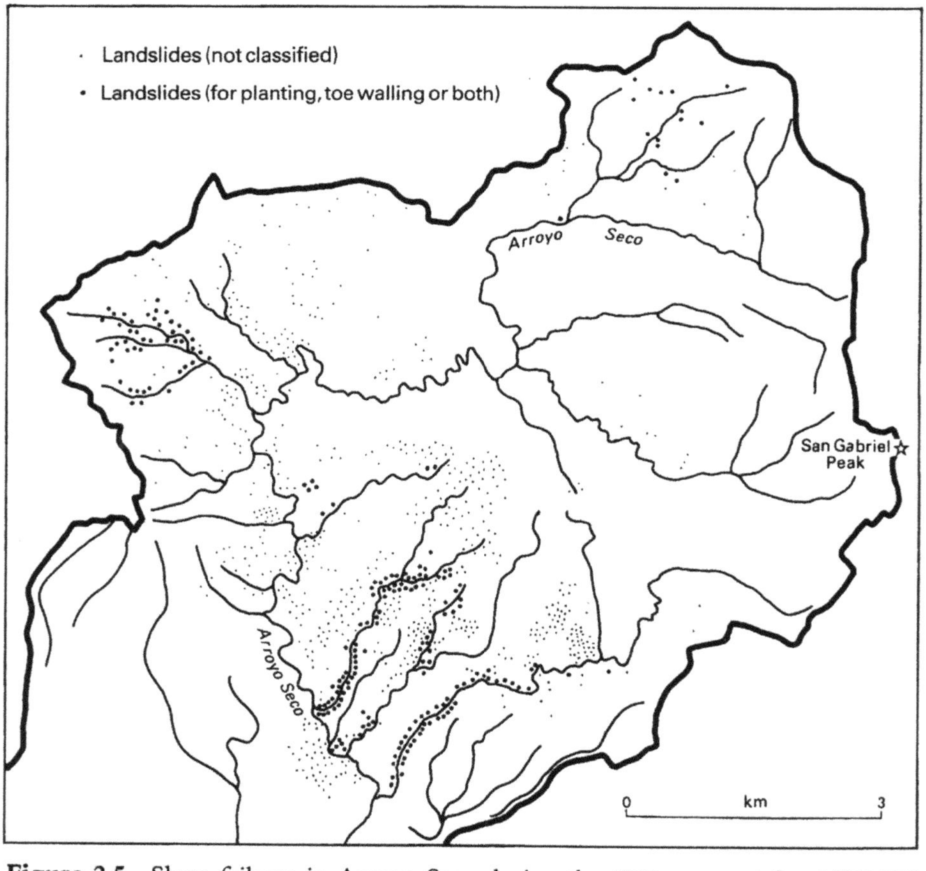

Figure 2.5 Slope failures in Arroyo Seco during the 1938 storms (after USDAFS 1939).

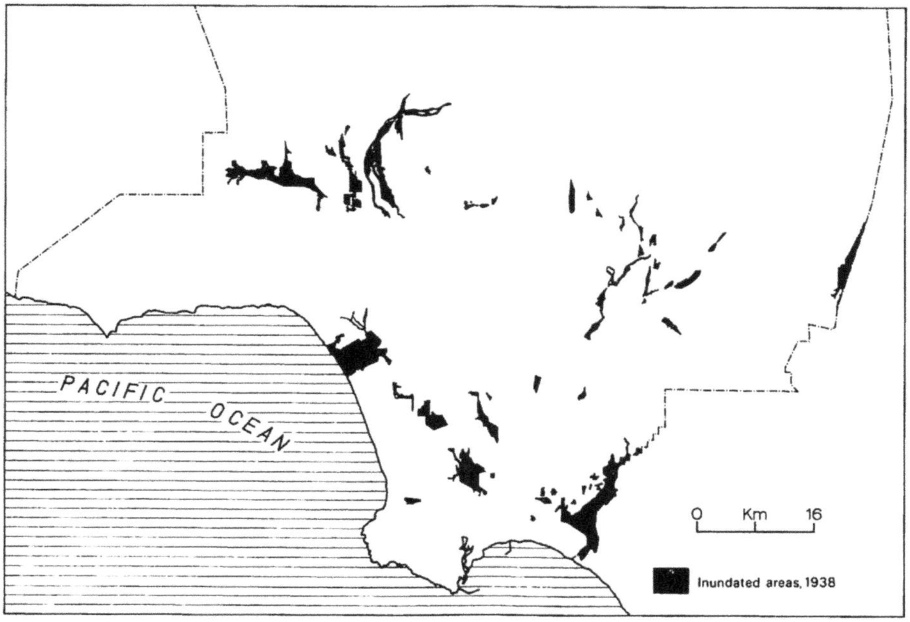

Figure 2.6 Approximate areas flooded in the San Gabriel and Los Angeles river watersheds in the 1938 storms (after LACFCD 1938).

the design flood capacity of reservoirs, with peak flows of over 7 m^3/s per km^2 being recorded: as a result, several reservoirs filled and overflowed, indicating the inadequacy of such control measures at the time (e.g. LACFCD 1938). Discharge in the Los Angeles River was the highest ever recorded. For example, 1.37 km upstream from Main Street Bridge, Los Angeles, 1896 m^3/s were recorded, compared with 622.6 m^3/s in 1934, and the previous record of 877.3 m^3/s in 1914. This was despite some reservoir storage added since 1914. The high discharge was due in part, of course, to the exceptional rainfall; but it was also the result of the effects of artificial flow concentration in man-modified channels, and of the increase in runoff caused by the 'urban effect' of the greatly expanded metropolis. As a result, over 43 710 ha were inundated (Fig. 2.6).

The floods were debris laden, although no debris flows like those of 1934 were reported. It is true that there had been a long period of debris accumulation in the mountains (as described above). But in many areas, the formerly anastomosing drainage networks across alluvial fans had been simplified into confining channels which transferred much debris downstream into the major river valleys, whereas it would have formerly come to rest on the fans. And the 16 debris basins and 13 reservoirs trapped much of the sediment. Loss of storage capacity in reservoirs averaged 12.5%, with over 15×10^6 m^3 of sediment deposited. Many of the 16 debris basins were largely filled, trapping some 527 052 m^3 of sediment (unreferenced article, 1938, LACFCD library). Some of the debris basins were overtopped.

Rates of erosion in areas tributary to reservoirs ranged from 3541 m^3/km^2 to 23 607 m^3/km^2, and averaged 15 049 m^3/km^2 (LACFCD 1938). Erosion rates in the smaller, steeper catchments tributary to debris basins were somewhat higher, ranging from 3836 m^3/km^2 to 35 411 m^3/km^2, and averaging 17 705 m^3/km^2 (LACFCD 1938). (These figures differ slightly from those in Troxell *et al.* 1942.) Debris basins constructed to protect the area flooded in 1934 (Dunsmuir, Eagle-Goss, Hall-Beckley, Hay, Pickens, and Shields basins) trapped over 413 324 m^3, and probably prevented serious damage in the La Crescenta–Montrose region. As a result of debris trapping, the debris problem in Los Angeles County was certainly less serious than it might have been; as a consequence, however, clear-out costs were very high. In general, debris accumulations were proportional to the slopes of the drainage catchments, although the new debris basins in the La Cañada area may have had relatively less debris than might have been expected because so much had already been removed in 1934 and the vegetative mantle was recovering.

Flood flows through the major valleys clearly carried much sediment from the mountains, but they were also highly erosive and accumulated debris from bank erosion on the plains. Debris was deposited in overflow areas, but large quantities were also carried to the sea. For example, the US Engineer Office (USACE 1938) estimated that some 5 042 400 m^3 were deposited at the mouth of the Los Angeles River.

Damage data are mainly provided for 'downstream' areas, by the USACE (1938). For the whole of the flood-affected area in southern California, 87 lives were lost, including 50 in Los Angeles County (of which 43 were in the Los Angeles River watershed). Total damage was estimated at $78 602 000, of which $27 413 000 was indirect (delays, breaks in communication, lost wages and business etc.), and $51 189 000 was direct (e.g. damage to property, repair, replacement and clean up). Of the downstream damage in Los Angeles County, which represented 40% of total storm damage ($31 522 000; the US Department of Agriculture (no date) estimated total damage in the Los Angeles River watershed as $37 693 000), the great majority was along the Los Angeles River, which burst its banks in the San Fernando Valley and further downstream in Los Angeles County, causing enormous damage to both agricultural and urban land (Table 2.5). The lower San Gabriel River valley suffered little damage because the river flowed via the Rio Hondo into the Los Angeles River.

Losses in the mountain areas are less well recorded. Troxell *et al.* (1942) give a few examples: $89 000 damage to government-owned recreational facilities; $12 550 damage to 24.6 km of telephone lines; the destruction of 520 cabins in the Angeles National Forest. The USDA (1938) estimated damage at $437 000 in the Arroyo Seco area, and the Los Angeles County Department of Forester and Fire Warden (1938) estimated $914 662 damage to fire-protection facilities

Table 2.5 Damage in Los Angeles County caused by storm of 27 February to 4 March 1938 (Troxell *et al*. 1942).

Drainage basin	Drainage area (km^2)	Damaged area (ha)	Direct damage ($) Recreational	Business	Industrial	Agricultural	Roads, bridges	Railroads	Utilities	Other	Total
Coyote Creek (San Gabriel R.)	469	24 666	899 000	518 000	30 000	1 446 000	172 000	106 000	125 000	65 000	3 361 000
San Gabriel River, except Coyote Creek	1 336	2 524	183 000	137 000	39 000	243 000	367 000	116 000	245 000	975 000	2 305 000
Los Angeles River, except Rio Hondo	2 027	10 193	2 186 000	1 020 000	70 000	447 000	6 135 000	511 000	1 952 000	4 675 000	16 996 000
Rio Hondo	365	2 537	360 000	137 000	61 000	176 000	240 000	72 000	180 000	483 000	1 709 000
Ballona and Topanga Creeks and miscellaneous coastal streams in Los Angeles County	1 271	3 789	369 000	249 000	15 000	63 000	850 000	none	1 593 000	31 000	3 170 000
Calleguas Creek	590	—	1 000	3 000	none	141 000	116 000	3 000	none	none	264 000
Santa Clara River	4 137	2 020	45 000	40 000	189 000	1 264 000	1 254 000	605 000	90 000	71 000	3 558 000

in the mountains of the county. These figures are but examples; actual losses must have been considerably higher.

The floods of 1938 occurred during the first major regional storms – storms that fully activated the whole fluvial system – since 1914. They were amongst the largest storms on record, and they caused the highest discharges ever recorded, mainly because of the antecedent land-use changes, especially urbanisation, that had taken place. Their impact was increased by the effect of the preceding wet period in February, but it was moderated by the effects of reservoirs and debris basins. The floods showed that debris yields could be as high from unburned as from burned areas. They demonstrated the indequacy of control measures or, to put it another way, the extent to which urban and rural development were proceeding faster than the introduction of flood controls. They also showed that a permanent disaster management organisation was needed, and that reliance on volunteers was outmoded; plans for disaster control and emergency organisation began to emerge after 1938 (Turhollow 1975).

The greatly increased awareness of the flood and water conservation problems of the region had led to an enormous improvement in data collection and monitoring facilities by 1938. As a result, the events were well recorded, and they provided an excellent basis for the re-appraisal of flood-control programmes. The time had come to secure the land and people against flooding in the same way that water supplies had been secured: by massive investment in engineering works. In this way, the two major prerequisites for urban growth – safe land and abundant water – were established.

But the problems were still essentially downstream problems, even though their cause may have been upstream. While damage in the hills and mountains in 1938 was reported, it was essentially to federal property and attracted little publicity. By 1969, there had been a dramatic change.

1969

In terms of their physical character and intensity, the storms of 1969 were similar in many ways to those of 1938. And yet the pattern of damage to both slopes and drainage systems was very different. Thus the storms of 1969 provide an excellent standard of comparison with 1938, and a measure of the dramatic changes that had occurred in Los Angeles County in the intervening 30 years.

The data available for this task are extensive, and for the first time the evidence of slope damage is almost of as good quality as that for flooding. The 1969 storms were probably monitored, investigated and analysed more intensively than any other storm event in the region, before or since. Channel and floodplain evidence and rainfall data were collated and reviewed by Simpson (1969) in his monumental report for the LACFCD. The USACE (1969) and the USGS (Waananen 1969, Hughes & Waananen 1972) also prepared reports, emphasising their own contribution to storm management and providing details of damage. These surveys are especially valuable because they incorporate data from numerous cooperating agencies. There are many other reports related to the storms (e.g. Rantz 1970, Scott 1971), together with volumes of unpublished material in the LACFCD library including information from municipal and other agencies.

But the main reports do not provide details of slope failure and erosion, problems of paramount importance in 1969 that were scarcely mentioned in 1938. The data on these problems are scattered in several locations, and have not previously been integrated. Indeed, some of the original data appear to have been lost or are no longer available. In the City of Los Angeles, the Department of Building and Safety is responsible for the collection of 'storm and slope failure damage reports' relating to private property. With the agreement of the department, a random sample of 20% of these reports has been analysed to help evaluate the impact of storms on slope problems. Other damage survey reports in the City of Los Angeles include those of the Engineering Geology section of the Engineering Department, and these, too, have been consulted. Various county agencies also collected data, many of which were summarised in the county application to the federal government (dated 26 September 1969) on behalf of the county and contracting cities, for federal disaster assistance under Public Law 81–875. An attempt has been made

DOI: 10.4324/9781003168645-8

to synthesize some of this huge body of data. In addition to these primary sources from local, state and federal agencies, the *Los Angeles Times* reports and press-cutting files have been analysed. These newspaper accounts provide some chronological control on events, illustrate many major features, and discuss the damage and deaths related to the storms. The whole body of data provides a satisfactory basis for assessing the geomorphological implications of the events.

Important antecedents to the events of 1969 include, over the longer term, urban development, flood and slope control, and, in the shorter term, fire and precipitation. As Fig. 1.17 shows, and descriptions by Preston (1967), Cooke and Simmons (1966) and Nelson (1959) make clear, southern California experienced unprecedented urban growth after World War II. The population of the county grew spectacularly, rising from 2 450 000 in 1938 to about 7 000 000 in 1969. Much of the urban development took place on land safe from flooding and slope failure, but some of it was in potentially hazardous locations, at times in defiance or accidental contravention of local planning and building controls. Of these hazardous developments, those in the hills and mountains are the most important, especially in the canyons and balmy slopes of the Santa Monica Mountains, the San Gabriel Mountains, and the hills of the Santa Clara River Basin. Development of some alluvial plains was also potentially dangerous, most notably where it was tributary to mountain canyons as yet uncontrolled by debris basins and flood reservoirs; in 1969 there were several such areas in the southern foothills of the San Gabriel Mountains, and it also transpired that many areas in the Santa Clara Valley and the Mojave Desert were inadequately protected. The urban development was not only sometimes in vulnerable locations, it also increased flood peaks and shortened flood lag times.

The storms of 1969 did not cause as much devastation and loss of life as the 1938 storms, largely because of the effectiveness of flood-control measures introduced to protect the most heavily populated areas since that time. By 1969, the comprehensive flood-control programme was almost complete – all 20 flood-control reservoirs had been built, 73 out of 106 planned debris basins were in operation, and some 2400 km of storm drains and 700 km of channel improvements had been completed. The success of local control measures – such as drain and channel improvements – is reflected in several city damage reports. For example, the city engineer of Compton stated that 'the city experiences during this last storm was [sic] a pleasant contrast with the problems which had been experienced in the past, prior to the construction of channels and storm drains through the city of Compton' (LACFCD, Bond issue study, contact information, 19 February 1969). The USACE (1969) estimated that the combined measures of the LACFCD, the USACE and the USDA *prevented* damage in 1969 to the value of $1013 million. (This figure appears to assume that developments up to 1969 would have taken place even if no flood-control measures had been implemented – an unlikely assumption –

but it more than justifies the total flood-control investment, which was of over \$1000 million by 1969.) In general, the flood-control system was well prepared for the storms, except for a few debris basins that already contained large quantities of material and were therefore unable to store much during the storms. Major flood problems occurred in inadequately protected areas, especially in the Santa Clara Valley drainage and the northern desert (which is beyond the LACFCD boundary; see Fig. 2.1).

Also, since 1938, urban development controls on sloping ground had been introduced to help prevent slope erosion and failure associated with houses, roads and other urban features built in the hills and mountains. The nature of these controls is considered in detail in Part 3 of this book, but it should be mentioned here that by 1969 a whole series of measures had been adopted, usually in response to particular slope failures. The events of 1969 provided a good opportunity to check the effectiveness of these and many other similar controls in the region.

The storms of 1969 were immediately preceded by several major fires, and by the hot, dry summer of 1968. Fires in five areas were important in leading to accelerated sediment yield the following year: the Easley Canyon fire (250 ha) and the Canyon Inn fire (730 ha) in the San Gabriel Mountains behind Glendora and Azusa; the Limerock fire, which burned 700 ha in the Little Tujunga catchment and 420 ha in the Pacoima Canyon catchment; and small fires in the hills behind Monrovia. Furthermore, rainfall before the storms was well below normal. For example, up to 17 January, precipitation at Opid's Camp was 51% of normal, and that in downtown Los Angeles 58% of normal (Simpson 1969). As a result, when the storms came, their initial impact was reduced because the ground soaked up the preliminary precipitation.

The storm sequence was more complex in 1968 than in 1938. It comprised a two-phase storm period from January 18–28, and a second complex but shorter stormy period from 22–25 February. The storms were not adequately predicted, a failure attributed to errors in the numerical forecasting models used (Bonner *et al.* 1971).

The autumn drought broke with rainfall on 13–14 January. Later, a large low-pressure system moved into southern California from the south-west bringing warm, moist air which mixed with cold Arctic air moving south (e.g. Simpson 1969, Bonner *et al.* 1971). The frontal system passed inland on 21 January, moving from north-west to south-east. It was followed by a ridge of high pressure that produced a lull in rainfall. A second frontal system moved into the area and reached Los Angeles late on 23 January, and stayed, trapped in the basin, until 25 January, bringing the heaviest rain. Full details of the meteorology of the storms are given by Bonner *et al.* (1971).

Throughout this and subsequent rainy spells, rainfall generally increased with altitude, and possibly as much as half of all precipitation arose from frictional convergence and lifting due to topography (Bonner *et al.* 1971). Storm totals for 18–26 January in the San Gabriel Mountains exceeded all

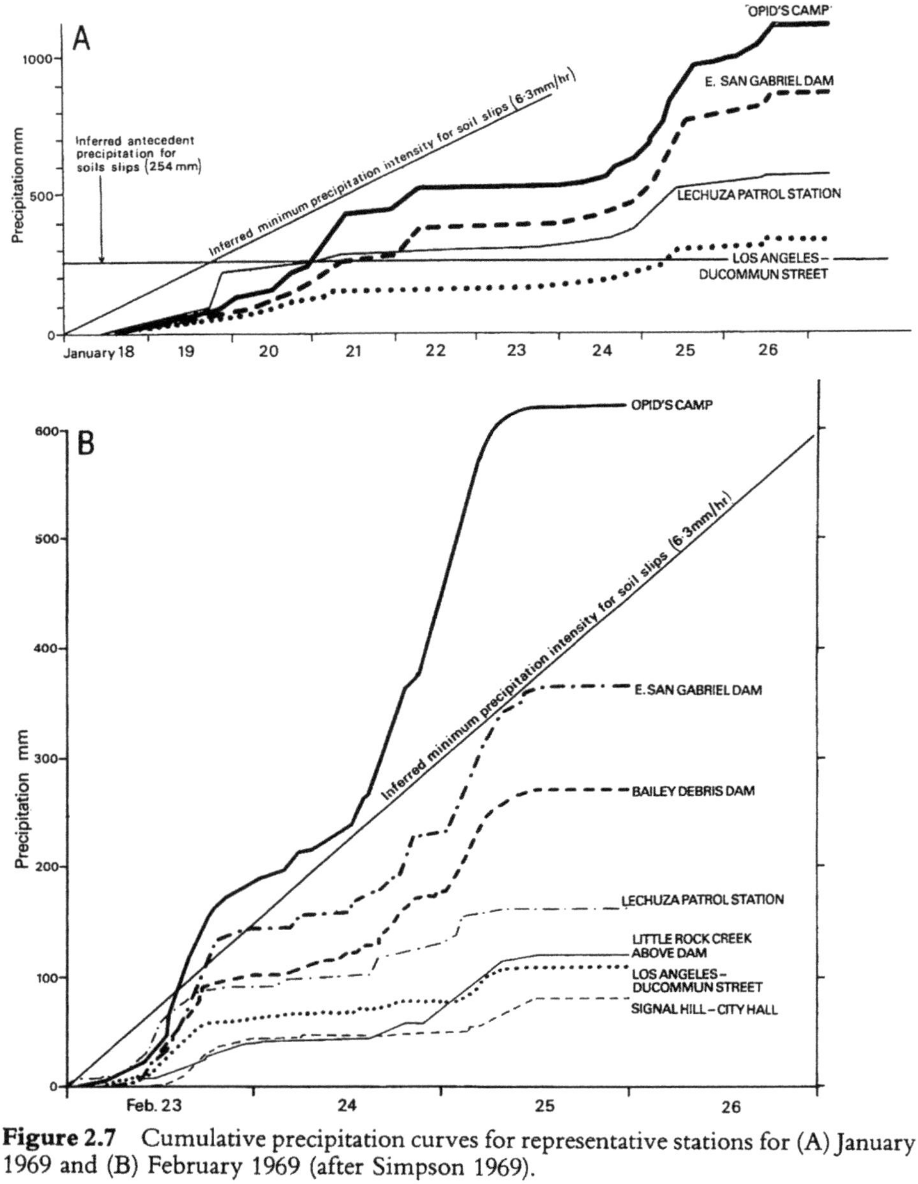

Figure 2.7 Cumulative precipitation curves for representative stations for (A) January 1969 and (B) February 1969 (after Simpson 1969).

previously recorded maxima. For example, Opid's Camp (altitude 1294 m) received 1145 mm, and San Gabriel Dam (altitude 709 m) received 868.7 mm; at Los Angeles (Ducommun Street, altitude 91 m), the total of 337.8 mm was also a record. The maximum 24-hour totals were also exceeded in the second phase of the storm at some of the more recently established recording stations.

Rainfall intensities in the January storms were generally lower in the first phase than in the second phase. At East San Gabriel Dam, for example, the

1-hour and 2-hour intensities rose to 38.1 mm and 63.5 mm respectively (compared with 30.5 mm and 58.4 mm in 1943, the previous highest records). In general, intensities were locally unusually high as, for example, the 43.2 mm/h at the Lechuza Patrol Station (Point Dume), and the 33 mm/h in the fire-ravaged hills above Glendora on 22 February. As Figure 2.7 shows, the intensities were greatest – and generally exceeded the approximate soil slippage threshold (the 6.35 mm/h intensity following a total of 254 mm) – in the afternoon of 24 January and for the 24-hour period from midday on 24 January to 25 January (Campbell 1975; see also Fig. 1.20). The recurrence interval of storm intensity varied between 1 and 10 years in the first phase, and 10 and 18 years in the second phase.

Rain fell intermittently throughout February, but major storms developed only on 22–25 February. A low-pressure system over the Gulf of Alaska moved south-east, the front passing Los Angeles on 23 February causing heavy rainfall in the mountains. The trough remained steady, and cold, unstable moisture-laden air caused prolonged precipitation when a second front moved in on 25 February. Precipitation was influenced by a series of waves associated with frontal passages from the southeasterly moving system and reached 152.4 mm in the mountains on 23 February, and 342.9 mm in 24 hours on 24–25 February at Opid's Camp, a maximum exceeding all previous February 24-hour maxima. Rainfall intensities were not record breaking, but they crossed the 6.35 mm/h assumed soil-slip threshold on 23 and 25 February (Fig. 2.7). The February storms brought particularly heavy rainfall to the north-west of the county.

Everywhere the February storms were less severe than those in January. They were nevertheless associated with serious damage, primarily because the ground was still largely saturated from the January storms. Their intensity had a recurrence interval of some 5–10 years. Overall, the storms were comparable to those in 1938, and larger than most other events (as Table 2.6 shows).

The pattern of runoff in 1969 was quite different from that of 1938, mainly because of the extent to which it was controlled. In general, however, inflow into reservoirs was in phases similar to, but somewhat after, the main phases of precipitation (Simpson 1969). Early in the period, rainfall tended to be absorbed by the dry ground, decreasing the initial effect of the storm, except in Malibu and the Santa Monica Mountains where there were high flows on 19 January, and in the Glendora–Azusa area on 22 January (see below).

Peak discharges did not generally occur until the latter part of the second phase, on 25 January. The long duration of the storm period resulted in very high total runoff which filled most of the LACFCD's reservoirs to spillway level, and used up much of the storage capacity of the USACE's flood regulation basins. In general, and as a result of these controls, peak flows were less than those in 1938. There were exceptions: for example, peak flow in the Los Angeles River below Wardlow Road (Long Beach) was 2886 m^3/s; in Topanga Creek, 345 m^3/s; and in the Santa Clara River, 418 m^3/s – all records.

Table 2.6 Rainfall comparisons (mm), 1914–69 (Simpson 1969).

	23–25/3/69	18–26/1/69	21–23/1/43	27/2–3/4/38	29/12–1/2/34	14–19/1/16	18–21/2/14
Opid's Camp							
max. 24 h	340.4	393.7	558.8[a]	406.4	340.4		
Total	534	1163.3[a]	856.0	693.4	454.7		
San Gabriel Dam							
max. 24 h	182.9	348.0	452.1[a]	264.2	193.0	—	190.5
total	327.7	868.7[a]	609.6	502.9	337.8	—	360.7
Glendora							
max. 24 h	96.5	152.4	137.2	170.2	208.3[a]	147.3	170.2
total	182.9	447.0[a]	284.5	348.0	411.5	355.6	330.2
Los Angeles							
max. 24 h	71.1	114.3	137.2	160.0	188.0[a]	106.7	—
total	99.1	337.8[a]	193.0	279.4	210.8	175.3	177.8

[a]Maximum record up to 1969.

Peak flows in February did not usually exceed those in January, although records were established in the north-west of the county, in Big Tujunga Wash, Verdugo Wash, Walnut Creek, and especially in the Santa Clara River (where 899 m^3/s was recorded). Peak flows on upstream tributaries of the Los Angeles River also exceeded January peak discharges. These high flows, and the high ratio of inflow to precipitation in this period, were undoubtedly related to the high antecedent precipitation that had saturated the surface.

FLOODING AND DEBRIS MOVEMENT IN CHANNEL SYSTEMS

The major locations of flooding during the 1969 storms are shown in Figure 2.8, which is based on data collected by the USACE and the LACFCD and other manuscript sources. Flooding was probably most severe in the headwaters of the San Gabriel and Los Angeles rivers, but it caused little damage to property, and was therefore not recorded in detail except in the thorough investigation of damage to concrete crib structures and metal binwall structures in Santa Anita and Sawpit canyons (LACFCD 1969). Downstream in the two major drainage systems, the established flood control system largely contained the flows, although 11 of the 14 LACFCD's reservoirs were overtopped on 25 January with uncontrolled spillway flow. Nevertheless, flooding of areas adjacent to channels was limited. Pumping station problems caused downstream flooding in Manhattan Beach, El Segundo, and Long Beach.

Undoubtedly, the more serious problems in the major drainage systems were associated with the floods and debris movement in the burned watersheds behind Glendora and Azusa in the San Gabriel Valley – events reminiscent of the La Cañada floods of 1934. Fire-promoted floods were anticipated here. For example, the Los Angeles County Fire Department (1969) organised meetings after the fires with the LACFCD, other agencies, and citizens of Glendora and associated vulnerable unincorporated areas, in order to advise them on the best methods of protecting property. Fire stations acquired thousands of sandbags from the LACFCD, and by November 1968 the Public Works Department of the City of Glendora had purchased 20 000 sandbags (Los Angeles County Fire Department 1969). And, in the autumn of 1968, an attempt was made to re-seed the fire-damaged watershed from the air (Los Angeles County Board of Supervisor's orders 13921 and 14188). But these preparations had only a limited effect. The first debris flows accompanied a particularly intense rainy spell on 21 January and early on 22 January. These were followed by others in different localities up to 25 January. The flows were highly charged with debris, and they tended to block channels and therefore become diverted on to adjacent alluvial fan slopes, where they flowed generally down streets, causing severe damage to adjacent property (City of Glendora 1969, Giessner and Price 1971). Such was the case, for example, at Sierra Madre Avenue, and most seriously in Glencoe Canyon (Glendora) on 25 January. Other affected canyons included Harding,

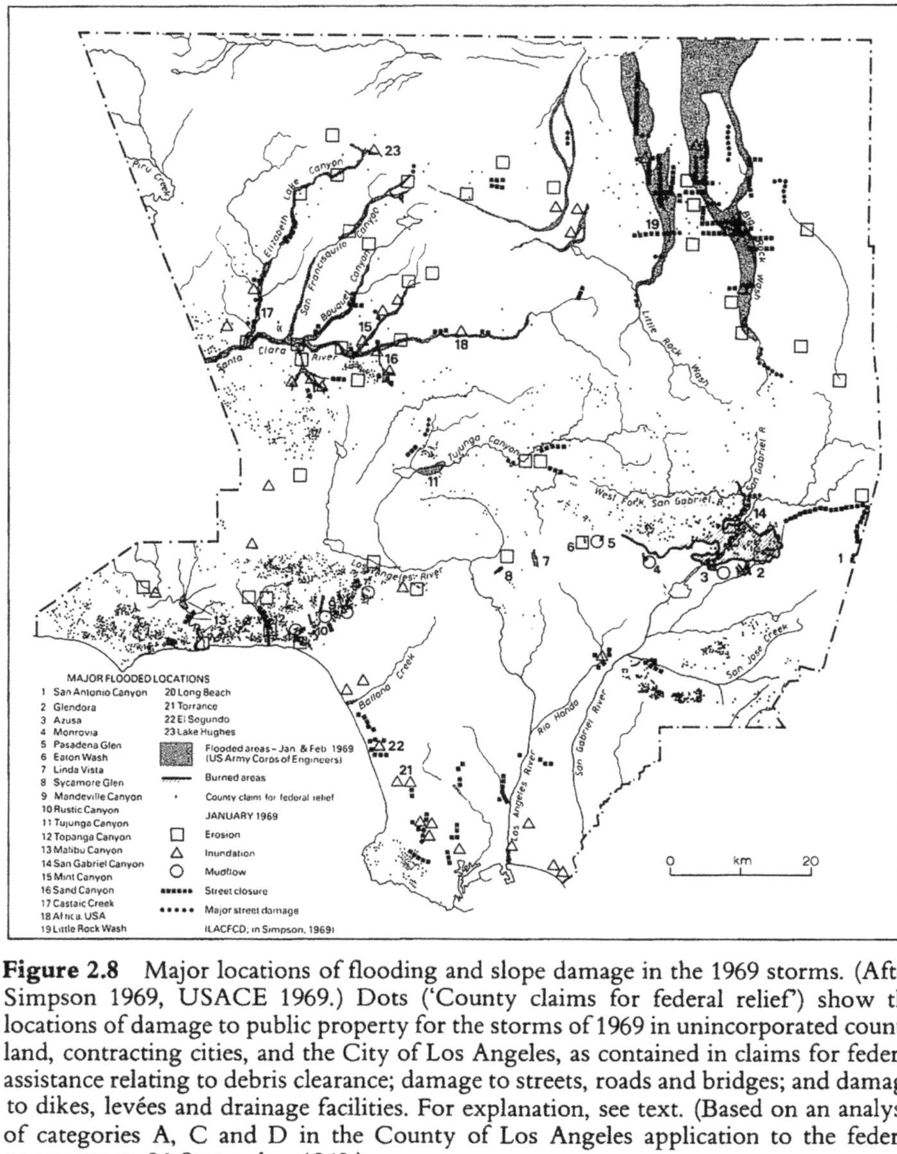

Figure 2.8 Major locations of flooding and slope damage in the 1969 storms. (After Simpson 1969, USACE 1969.) Dots ('County claims for federal relief') show the locations of damage to public property for the storms of 1969 in unincorporated county land, contracting cities, and the City of Los Angeles, as contained in claims for federal assistance relating to debris clearance; damage to streets, roads and bridges; and damage to dikes, levées and drainage facilities. For explanation, see text. (Based on an analysis of categories A, C and D in the County of Los Angeles application to the federal government, 26 September 1969.)

Pennsylvania, Rainbow, Gulscourt, Hook, Hicrest, Beatty, and Hilltop. Burial of houses without their destruction (a feature of early phases of debris flows) occurred in places, such as Rainbow Drive (Glendora); most damage accompanied more fluid flows. In Hook Canyon, one permanent and a temporary debris basin helped to delay debris flows, but the former was filled and overtopped and the latter damaged, so that a serious mainstream flow developed from an accumulation of tributary flows. Additional damage occurred in Glendora on 7 February (*Los Angeles Times*, 13 February 1969). Total damage to public and private property was $2 500 000 in Glendora and Azusa. Minor debris flows were associated with burned areas behind Monrovia, and also in Sycamore Canyon (Glendale), and Pasadena Glen (Sierra Madre).

Further geomorphological details of these localised alluvial-fan events have been described by Scott (1971) and by Giessner and Price (1971). Their recurrence interval, as judged by historical observations at nearby recording stations, was in the order of 70 years.

The 'fanhead valley' of Tujunga Wash also suffered serious damage. This locality was analysed in detail by Scott (1973) using air-photo-based design maps made before and after the storms. Scott showed that the unusually high discharge of water and sediment down Tujunga Wash led to hydrological adjustments in the alluvial valley floor that are similar to those expected on alluvial fans. There was extensive net scour (up to 6.1 m ± 0.6 m) and net fill (up to 10.67 m ± 0.6 m); unexpected, but natural, diversion of flow from one distributary channel to another radiating from the fan apex, perhaps arising from local deposition, caused inundation and damage in the Westcott Avenue area; lateral erosion in channels adjusting to unusually high discharges caused part of an old alluvial terrace to be destroyed, taking much of Bengal Street with it; and headward erosion consequent upon the lowering of base level as the flood entered a large gravel pit caused the failure of three bridges. Clearly, urban development in this area had taken place without a full recognition of the potential hazards.

Debris generation in the Los Angeles and San Gabriel river watersheds reached record levels in January. Sediment yields in the burned areas were exceptionally high, with a record accumulation, in East Hook Debris Basin of 41 313 m^3/km^2, and 36 001 m^3/km^2 in Harrow Debris Basin, both in the Glendora–Azusa area. But unburned areas also showed high yields. For

Table 2.7 The storms of 1969: summary of sediment yield (m^3/km^2) – Simpson (1969).

	January	February	Rain season
Catchments tributary to reservoirs (14)			
minimum	2 632	1 088	1 088
maximum	22 692	8 498	22 692
average	9 833	3 442	6 662
Catchments tributary to debris basins (73)			
minimum	0	0	0
maximum	41 313	24 492	41 313
average	8 896	3 895	6 369

example, Big Dalton Debris Basin had a yield of 25 555 m^3/km^2, and the West Kinneloa Debris Basin, a rate of 31 280 m^3/km^2. In February, yields were lower (Table 2.7). For example, the Harrow Basin received only 7436 m^3/km^2, and the Big Dalton Basin, 20 322 m^3/km^2. This reduction reflects the lower intensity of the February storms, and the fact that much of the available debris was removed by the January storms. For the season as a whole (Table 2.7) the debris yields were somewhat lower than those for 1938.

Despite substantial sediment yields in both the January and February storms, and the local occurrence of debris flows, most sediment was contained either in stabilisation structures constructed by the USDA in the mountains, or in debris basins and reservoirs operated by the LACFCD. The mountain stabilisation structures were largely filled in the January storms, and a comparison of 'stabilised' with 'unstabilised' watersheds showed that, in the areas of highest erosion, yields in the former were reduced by about 40% (Simpson 1969). In the February storms, these structures, although filled, still reduced debris yield by about 16%, probably because the earlier accumulations tended to stabilise channel banks and foot slopes, thus reducing debris supply. Of the debris basins, 7 were filled and overtopped (including those in the Glendora–Azusa region), and 42 were filled to over 25% of their design capacity in January. By February, many debris basins had been at least partially cleared by the LACFCD and were able to accommodate the February supply; none was filled to capacity in the later storms. The flood-storage reservoirs all had their capacity reduced in both January and February. For example, the San Gabriel Reservoir accommodated 2 535 716 m^3 in January and 2 502 100 m^3 in February, so that in January 72.8% of its original capacity remained, and by March this had been reduced to 68.9%. In total, the LACFCD's reservoirs accumulated 7 586 520 m^3 in January, and 4 844 524 m^3 in February; the debris basins accumulated 1 506 559 m^3 in January and 450 760 m^3 in February.

Flooding elsewhere during the January storms occurred mainly in the west and north of the county. In the Santa Monica Mountains, floods caused considerable damage in Malibu and Topanga canyons, and to a lesser extent in Rustic and Mandeville canyons. In the Santa Clara Valley and in the Antelope Valley (principally Little Rock and Big Rock creeks) – both areas with less experience of floods and less well protected – flooding was extensive. The Santa Clara, Big Tujunga and Antelope Valley catchments were the principal sites of flooding and damage to property, bridges and roads in the February storms, and in all cases the damage was more severe than previously. In the region as a whole, damage from February storms was lower, but in both periods it was highest in the basin of the Santa Clara River. But it is clear that, by comparison with 1938 and 1914, the areas inundated were small and scattered, and in general located beyond the catchments that had been the primary concern of the LACFCD and the USACE. Most damage was caused by flooding in areas of new development. Once again development had progressed more rapidly than hazard control.

EROSION AND FAILURE ON SLOPES

Data on damage caused by slope erosion and failure during the 1969 floods were collected by many agencies. Unfortunately, the data are often unclear in defining the nature of damage in geomorphological terms, and it has proved impossible to assemble a comprehensive map of slope damage in the county from the incomplete data available. The closest approximation to such a map is based on two sources of data: the Los Angeles County submission to the federal government for assistance under Public Law 875 (81st Congress), and the storm and slope-failure damage reports of the City of Los Angeles' Department of Building and Safety.

The county's federal submission, dated 26 September 1969 (one of several documents prepared by the county), provides a list of claims for damage arising from the whole storm season as it affected county agencies; some data for the City of Los Angeles were also attached to the report and have been included in this analysis.

The locations of damage have been plotted in Figure 2.8 (p. 86) for three categories of claims – debris clearance, damage to streets, roads and bridges, and damage to dikes, levées and drainage facilities. The documents in the county's federal submission exclude claims relevant to the LACFCD and USACE, and the copies available to the author are incomplete (e.g. some pages are missing). From them it is impossible to determine which claims relate to slope erosion and failure. Nevertheless, their distribution is complementary to that shown for flooding on Figure 2.8, and reveals that damage was very much more widespread than that described in the reports of the LACFCD and USACE. There was clearly much damage away from the main drainage channels, and it was concentrated in the Santa Monica Mountains, the Palos Verdes and Puente hills, the foothills of the San Gabriel Mountains (especially in the east), in the hills north of San Fernando Valley, and in the Antelope Valley of the Mojave Desert. Most of the few large claims concern such features as bridge collapse; small claims were predominant, and most of them involve debris clearance. The dot distribution on Figure 2.8 therefore probably provides a reasonable impression of the damage pattern arising from soil slips, landslides and erosion on slopes in the unincorporated land of the county, the cities for which the county was responsible, and the City of Los Angeles.

A complementary picture for private property in the City of Los Angeles emerges from an analysis of the storm and slope-failure damage reports of the city's Department of Building and Safety. These reports were made by the department's inspectors, largely in response to calls from the public, and each consists of a completed standard, coded form. The information recorded varies somewhat between reports, mainly because some categories of data are imprecisely defined and many different surveyors were involved. Nevertheless, the reports constitute as full and accurate a survey of damage to private properties as exists for 1969, and a 20% random sample of them is analysed in some detail below. It should be noted that the reports refer to

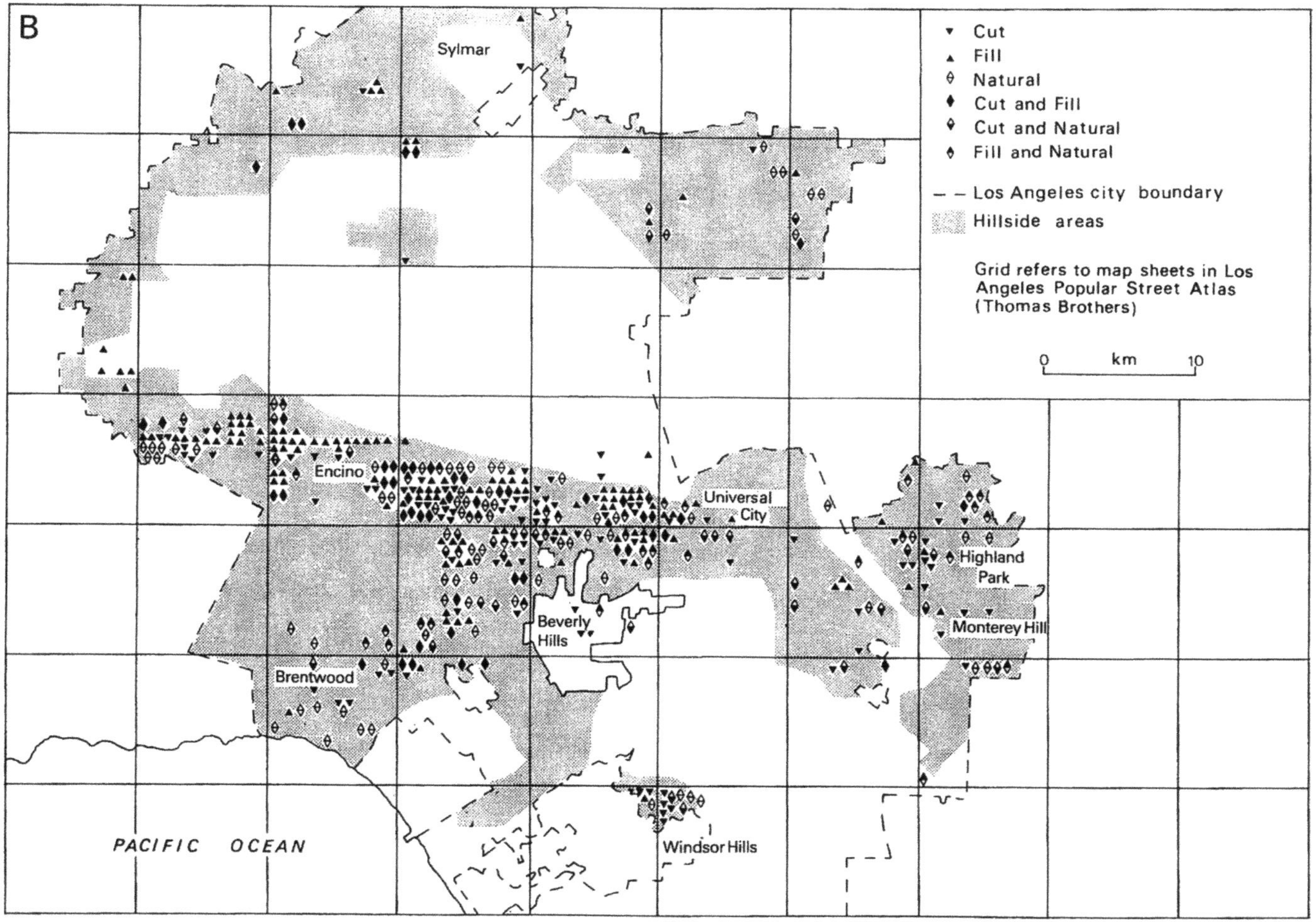
B
Sylmar
Cut
Fill
Natural
Cut and Fill
Cut and Natural
Fill and Natural
Los Angeles city boundary
Hillside areas
Grid refers to map sheets in Los Angeles Popular Street Atlas (Thomas Brothers)
0 km 10
Encino
Universal City
Highland Park
Monterey Hill
Beverly Hills
Brentwood
Windsor Hills
PACIFIC OCEAN

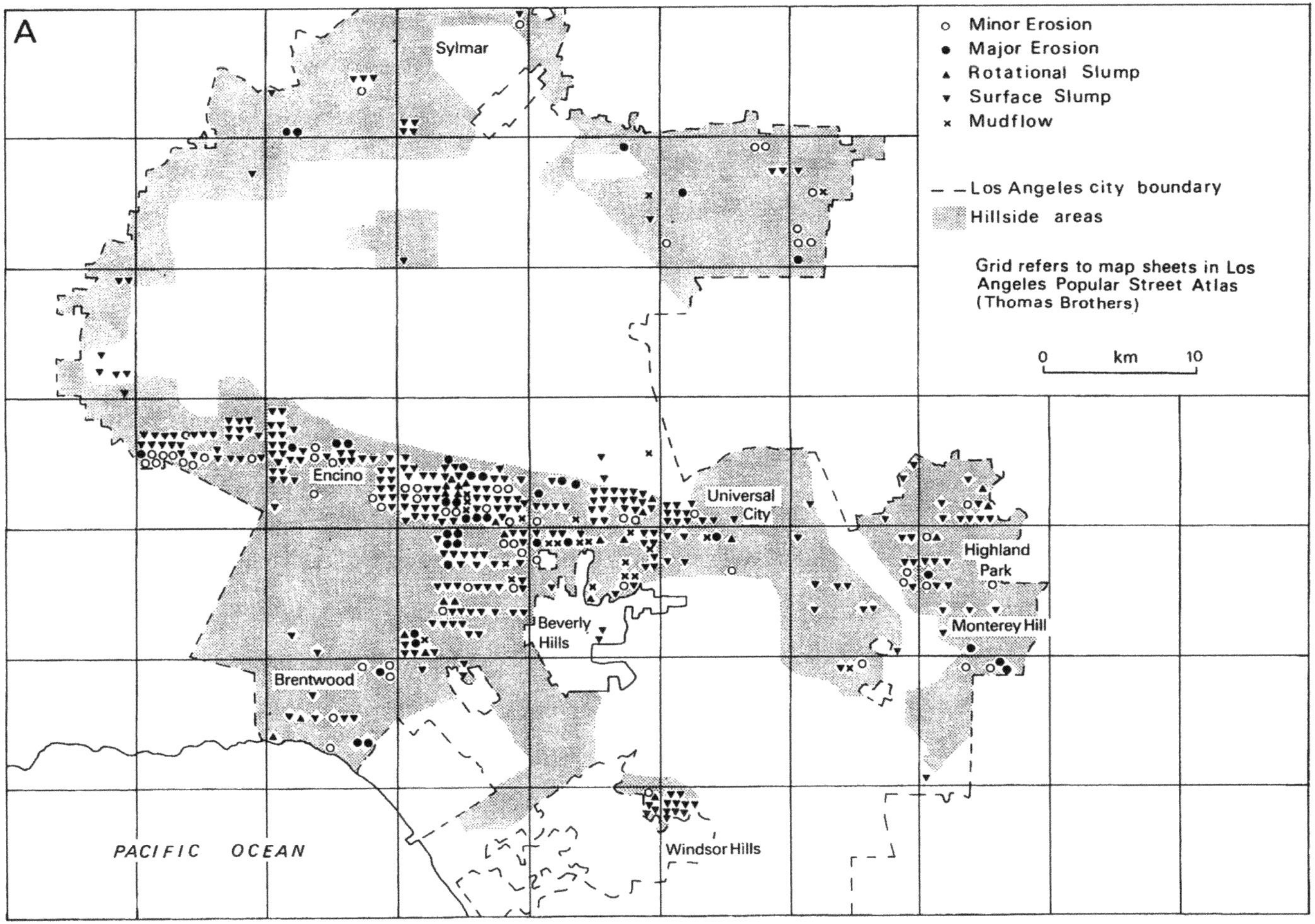
A
Minor Erosion
Major Erosion
Rotational Slump
Surface Slump
Mudflow
Los Angeles city boundary
Hillside areas
Grid refers to map sheets in Los Angeles Popular Street Atlas (Thomas Brothers)
0
km
10
Sylmar
Encino
Universal City
Highland Park
Monterey Hill
Beverly Hills
Brentwood
Windsor Hills
PACIFIC OCEAN

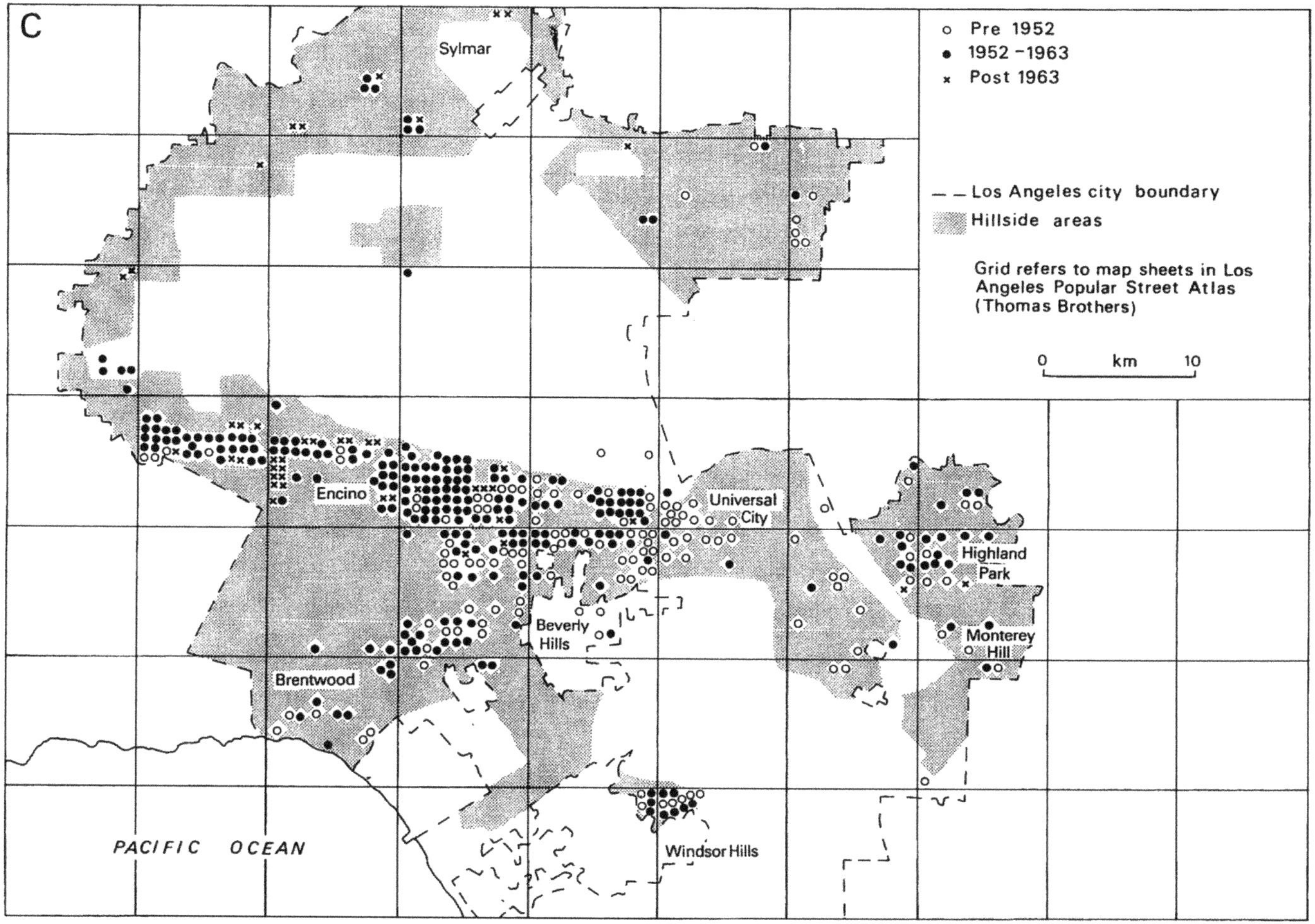

Figure 2.9 Spatial pattern of damage to private property within the City of Los Angeles, as revealed by an analysis of a 20% random sample of storm and slope-damage reports collected by the Department of Building and Safety. (A) Type of slope problem. (B) Types of slope affected. (C) Age of affected property.

individual properties; where a failure affects more than one property (as is often the case with large failures), each affected property will have a report.

The distribution of the sample of reports and failure types is shown in Figure 2.9A. Damage to private property is clearly concentrated in the hills and mountains; indeed, it is almost entirely within the 'hillside areas' designated by the city as potentially hazardous. The abundance of damage in the Santa Monica Mountains is outstanding, especially in the relatively new communities along the north flanks, such as Woodland Hills, Encino, Sherman Oaks, Tarzana, and Studio City. Failures were also numerous on the southern flanks of the mountains, for example in Brentwood, Bel-Air and behind Beverly Hills and Hollywood. Problems were also evident in the recent community developments in the Verdugo Hills, and in Highland Park (near Glendale).

The survey distinguishes different types of failure (Table 2.8a) and, although the distinctions amongst types may be arbitrary, it is clear that the great majority of failures were 'surface slumps' (i.e. soil slips), with minor soil erosion being second in importance (Fig. 2.9A). 'Mudflows' are slope debris

Table 2.8 The storms of 1969: slope-failure data in the City of Los Angeles. (All data are based on a 20% sample of storm and slope-failure damage reports prepared by the Department of Building and Safety, City of Los Angeles).

(a) Types of failure.

	Absolute frequency	Relative frequency (%)
major erosion	40	8.4
minor erosion	65	13.8
surface slump	294	61.5
rotation	21	4.4
mudflow	23	4.8
other	6	1.3
no data	28	5.9

(b) Slope conditions (up to five types of damage may be recorded for each of 477 sites).

	Number of occurrences	Percentage of 477 sites
berms removed, drainage over slope	71	14.9
drainage terrace blocked	26	5.5
unsupported cut at toe	24	5.0
slope saturation	416	87.2
evidence of water ponded at top	20	4.2
poor yard (garden) drainage	74	15.5
subsurface water	7	1.5
other	15	3.1
total	653	

Table 2.8—*cont.*

(c) Slope characteristics.

	Absolute frequency	Relative frequency (%)
cut	105	22.0
fill	139	29.0
natural	82	17.2
cut and fill	53	11.1
cut and natural	31	6.5
fill and natural	36	7.5
inadequate data	31	6.5

(d) Type of failure and slope characteristics (%).

	Erosion				
	Major	Minor	Mudflow	Rotation slump	Surface slump
cut	25.0	35.4	4.3	23.8	22.4
fill	22.5	9.2	17.4	9.5	39.5
natural	25.0	27.7	34.8	23.8	12.6
cut and fill	10.0	9.2	0	9.5	13.3
cut and natural	10.0	10.8	17.4	14.3	4.4
fill and natural	5.0	6.2	17.4	14.3	7.8

(e) Occurrence of failures in relation to soil type.

	Absolute frequency	Relative frequency (%)
crystalline rock	17	3.6
stratified sedimentary rock	48	10.0
soft and broken bedrock	83	17.4
sand (silty, clayey)	206	43.3
silt	3	0.6
clay (lean, fat, organic)	82	17.2
other	8	1.7
no data	30	6.3

(f) Slope type and slope inclination (% of each type per slope inclination category).

Inclination							
(degrees)	(%)	Cut	Fill	Natural	Cut and fill	Cut and natural	Fill and natural
<33	62	2.9	7.2	9.8	13.2	6.5	11.1
33	65	19.0	71.9	9.8	52.8	9.7	22.2
34–44	67–96	6.7	12.2	32.9	1.9	22.6	33.3
45	100	53.3.	4.3	35.4	22.6	25.8	13.9
>45	>100	17.1	2.9	9.8	9.4	32.3	16.7

Table 2.8—*cont.*

(g) Relation of slope-failure type to slope inclination (% of each failure type per inclination category).

Inclination		Erosion				
(degrees)	(%)	Major	Minor	Mudflow	Rotation slump	Surface slump
<33	62	2.5	9.2	4.3	4.8	8.5
33	65	30.0	20.0	13.0	38.1	44.2
34–44	67–96	12.5	15.4	34.8	19.0	14.6
45	100	27.5	40.0	21.7	14.3	22.8
>45	>100	20.0	13.8	17.4	14.3	8.8

(h) Vegetation cover at failure sites.

	Absolute frequency	Relative frequency (%)
no planting	29	6.1
ice plant, large leaf	41	8.8
ice plant, small leaf	26	5.6
ivy	68	14.2
ground planting cover general	46	9.6
combined ground cover, low shrubs and trees	78	16.3
grasses	74	15.5
native vegetation	80	16.5
other	4	0.8
no data	31	6.5

(i) Density of vegetation cover at failure sites.

	Absolute frequency	Relative frequency (%)
dense	187	39.1
spotty (dense in spots)	96	20.1
sparse	129	27.0
other	12	2.7
no data	53	10.7

(j) Dimensions of slope failures.

	Range (m)	Median (m)
maximum width of failure	0.91–121.92	7.62–9.14
maximum height of failure	0.91–30.48	6.10
maximum depth of failure	0.30–9.14	0.81

Table 2.8—*cont.*

(k) Conservation practice at failure sites.

	Absolute frequency	Relative frequency (%)
none	257	53.8
drain	20	4.2
terrace	34	7.1
buttress	2	0.4
terrace drain	51	10.9
retaining wall	81	16.5
terrace and wall	6	1.3

(l) Relation of type of slope failure to conservation practice (%).

	Minor erosion	Rotation slump	Surface slump
terrace and wall	0	0	2.0
retaining wall	16.9	19.0	15.3
terrace drain	3.1	9.5	13.9
buttress	1.5	0	0.3
terrace	10.8	9.5	6.8
drain	4.6	4.8	4.8

(m) Ordinances related to slope failures.

	Absolute frequency	Relative frequency (%)
pre-1952	132	27.6
1952–63	243	51.0
after 1963	46	9.6
no data	56	11.7
total	477	100.0

(n) Failure at sites conforming or not conforming to ordinances in force at time of site development.

	Absolute frequency	Relative frequency (%)
conforming	291	61.1
non-conforming	114	23.8
no data (or no failure)	72	15.1
total	477	100

flows (often associated with soil slips); they are not channel debris flows as described above. Not unexpectedly, the slope conditions at damage sites point overwhelmingly to slope saturation (whether due to high rainfall, blocking of drains, or ponding) as the major cause of failure (Table 2.8b).

The physical character of damage sites is dominated by both natural conditions, and by site changes imposed by developers. Thus, of the slopes affected, there are six different types (Fig. 2.9B, Table 2.8c), of which the majority are cut, fill or natural. Surface slumps were most common on fill slopes, rotational slumps and surface erosion are predominantly on cut and natural slopes, and most mudflows were on natural slopes (Table 2.8d). Materials underlying the failed slopes were predominantly 'sand' (silty, clayey) – mainly because this category includes fill and superficial sediments (Table 2.8e).

Slope inclination is also important. Nearly all failure reports relate to slopes of 33° or over (Table 2.8f), with a staggering 70.4% of cut-slope failures on slopes of 45° or more; 71.9% of fill-slope failure reports and 52.8% of cut-and-fill-slope failure reports concerned slopes at the pre-1963 ordinance limit of 33°. Most natural slopes that failed were between 34° and 45°. Similarly, most surface and rotational slumps were on slopes at the 33° limit (or higher); mudflows were largely on slopes in the 34°–45° range; and erosion, too, was clearly associated with steeper slopes (Table 2.8g).

Most failures occurred on vegetated slopes (Table 2.8h), mainly grass, 'native' vegetation, or a ground cover/shrub–tree combination. Because of the variety of vegetation types and density (Table 2.8h & i), it seems possible that vegetation did not play a fundamental rôle in determining failure location.

In general, failures were small, averaging 7.62–9.14 m wide, 6.1 m high and 0.8 m deep (Table 2.8j). A high proportion of damaged sites (53%) had no conservation (slope stabilisation) work on them (Table 2.8k). Of damage sites with conservation measures, terraces, terrace drains and retaining walls were the most frequently used, and they were primarily associated with surface slumps and erosion (Table 2.8l); major erosion and mudflows only occurred on slopes without conservation practices.

The data also allow an assessment of the effectiveness of building control ordinances in preventing slope failure. In the 477 reports analysed, some of the buildings affected were constructed before grading ordinances were introduced in 1952, some were built between 1952 and the 1963 ordinances, and some were built after 1963 (Table 2.8m). Clearly, older properties were most affected, for less than 10% of the sample was constructed after 1963 (Table 2.8m). Of the failures, over 20% did not conform to ordinances existing when they were built, but over 60% did conform (Table 2.8n).

There is a strong spatial (as well as temporal) differentiation to the pattern of damaged properties on slopes (Fig. 2.9C), with the older affected sites chiefly in the east of the city and in the east of the Santa Monica Mountains, and the younger sites mainly towards the west and north. It is interesting to compare this pattern of essentially superficial slope failures with that of the predicted landslide potential zones on Fig. 1.21 (p. 21), which is based only on natural variables. The zone of maximum slope damage in 1969 occurs in areas of low, medium and high predicted landslide potential, and the difference between the patterns probably reflects both the different processes at work and the injudicious human interference with the natural situation.

In order to answer the question 'How effective were the controls on building development?', sites that failed must be assessed as a proportion of all building sites developed. Although the number of buildings constructed in hillside areas at different times is not known precisely, estimates of construction sites for the city are given by Slosson (1969). Slosson showed that the predictable percentage failures in 1969 of sites constructed before 1952, between 1952 and 1963, and between 1963 and 1969 were 10.4, 1.3, and 0.15 respectively. This relationship between damage and the age of building sites presumably reflects the increasing effectiveness of the building controls and their implementation. Slosson (1969) also showed that the estimated average damage cost per site for the total number of sites developed fell from $330 for pre-1952 sites to only $7.0 for 1963–68 sites – a further convincing demonstration of the effectiveness of the building codes.

In the earlier discussion on soil slips, it was noted that failures seemed to be associated with times when total storm rainfall had reached 254 mm and subsequently rainfall intensity rose to 6.3 mm/h or more (Campbell 1975). In

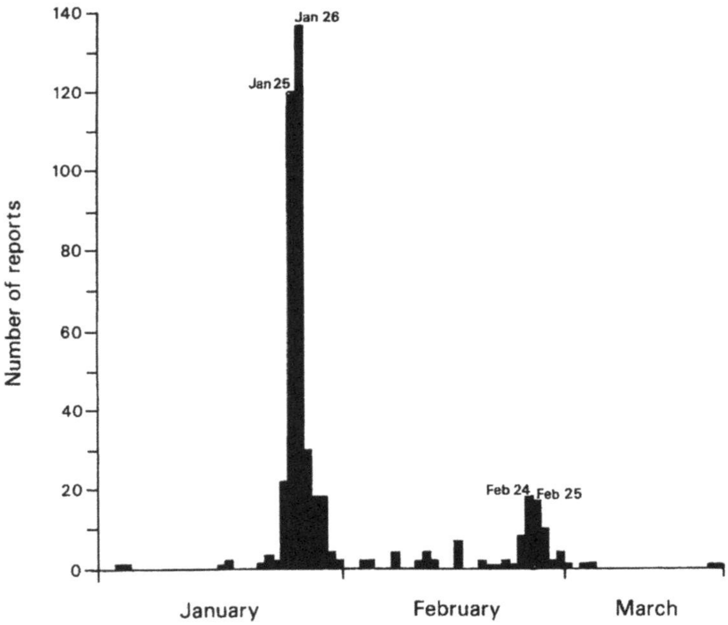

Figure 2.10 Dates of storm and slope-failure damage reports analysed, 1969. For explanation, see text.

1969 (as Campbell explained) the antecedent rainfall threshold was crossed in Los Angeles County (mainly between 19 and 21 January), and the intensity threshold was crossed thereafter on several occasions, including the morning of 22 January, 25 January, and at times in February. The date on which storm damage was first reported by the public is given on each report. Figure 2.10 shows that the 25–26 January and 24–25 February periods above the notional threshold were the times of most reports, but that few failures seem to have occurred in the first storm phase on 22 January, perhaps because ground saturation was inadequate to cause failure by that date. Over 77% of failures occurred in January, and only 20% in February/March.

The 1969 storms also generated some major landslides, principally in the notoriously unstable area of the Pacific Palisades (USACE 1969). This area comprises a series of deeply entrenched mesas and, along the coast, steep cliffs in relatively incoherent rocks. Earlier slides were mapped by McGill (1959 & 1973). In 1969, slides occurred in Revello Drive and Castellamare, in the Via de los Olas (i.e. in the bluffs overlooking the Pacific Coast Highway), together with the Via Santa Yñez. Total damage here was estimated at $1 119 000. Not all other large failures coincided with heavy rainfalls, as did most soil slips. Some were delayed days, even weeks. For example, the huge slope failure and debris slide on the Pomona Freeway occurred on 12 February (*Los Angeles Times*, 13 February).

There is much evidence to support the view of many of the environmental management agencies in Los Angeles County, that their measures to combat slope and channel problems were dramatically successful in diminishing the impact of the 1969 storms. Certainly flooding hazards were most severe in the areas beyond the extensive control measures in the Los Angeles and San Gabriel river watersheds, with the important exception of the Glendora–Azusa problems, which were in part due to the fact that sediment-control measures had not been completed. Certainly, too, slope failures particularly affected older properties, properties not subject to the latest controls on grading in the City of Los Angeles.

COUNTING THE COST

The impact of the 1969 storms in terms of loss of life, damage to property and cost, was enormous. Overall, estimated damage exceeded $81 million; at least 74 people were killed, and over 10 000 people had to evacuate their homes. The problems were on a scale that qualified for both state and federal relief. The state declared an emergency on 21 January (*Los Angeles Times*, 22 January), and a State of Disaster was declared in Los Angeles by Governor Reagan on 23 January (*Los Angeles Times*, 24 January). The state also passed flood relief acts to raise funds through a gasoline tax (e.g. *Los Angeles Times*, 20 March). On Sunday 26 January, President Nixon made an initial allocation of $3 million for relief, federal loans became available to homeowners and businesses, and householders in the region were able to claim tax relief for damage repair.

Several agencies prepared estimates of damage and repair costs, and summaries were assembled by the LACFCD (Simpson 1969) and the USACE (1969). The costs of flood damage alone are given in Tables 2.9 and 2.10. There are clearly some discrepancies between estimates in these tables. For example, Simpson's estimates of January damage greatly exceed those of the USACE, possibly because they include additional data the LACFCD received from many agencies and cities in January. The corps' estimates do not include the $16 million required for debris removal from LACFCD debris basins, or the cost of damage to hillslope property away from flooded areas. The data in Table 2.9 emphasise the importance of flood damage in the county outside of the two main drainage systems (69%), and Table 2.10 illustrates the relative importance of damage to highways and bridges, to public property, and to flood-control facilities.

Table 2.9a Total storm damage in Los Angeles County, 1969 (Simpson 1969).[a]

	January ($)	February ($)	Total ($)
public property	53 463 000	13 386 000	66 849 000
private property	14 194 000	645 000	14 839 000
total damage	67 657 000	14 031 000	81 688 000
number of lives lost	53	21	74[b]

[a]Excluding costs of debris removal.
[b]Includes at least 16 by drowning and 57 in storm- and flood-related traffic accidents (USACE 1969).

Table 2.9b Total flood damages in principal drainage areas, floods of January and February, 1969, Los Angeles County (Simpson 1969).

	Damages ($)[a]		
Drainage area	January 1969 flood	February 1969 flood	Total ($)
Santa Monica coastal streams drainage areas	1 714 000	198 000	1 912 000
I os Angeles River drainage areas	3 024 000	2 597 000	5 621 000
San Gabriel River drainage areas	3 212 000	825 000	4 037 000
Santa Ana River drainage areas	1 223 000	0	1 223 000
Antelope Valley drainage areas	735 000	1 477 000	2 212 000
Santa Clara River drainage areas	6 259 000	10 132 000	16 391 000
Total	16 167 000[a]	15 229 000[a]	31 396 000[a]

[a]Includes monies spent under Public Laws 84-99 and 81-875; does not include an estimate for removal of debris from debris basins by the LACFCD of $16 000 000.

Table 2.10 Flood damages along streams in Los Angeles County, 1969 (USACE 1969).

Type of property	Damages ($) January, 1969			Damages ($) February, 1969		
	Physical damages ($)	Business loss and emergency cost ($)	Total ($)	Physical damages ($)	Business loss and emergency cost ($)	Total ($)
residential	2 480 300	359 700	2 840 000	1 130 700	164 300	1 295 000
business	1 167 500	858 500	2 026 000	587 000	365 000	952 000
agricultural	256 500	56 500	313 000	508 000	103 000	611 000
highways and bridges	2 936 500	735 500	3 672 000	3 310 300	1 130 700	4 441 000
utilities	905 400	269 600	1 175 000	854 500	473 500	1 328 000
public	3 586 800	172 200	3 759 000	3 378 500	34 500	3 413 000
flood control (including channels)	1 236 000	1 000	1 237 000	1 325 000	0	1 325 000
railroad	65 000	23 000	88 000	588 000	496 000	1 084 000
undeveloped land	136 000	0	136 000	192 000	0	192 000
harbour	886 000	0	886 000	565 000	0	565 000
recreational	28 000	7 000	35 000	19 000	4 000	23 000
total	13 684 000	2 483 000	16 167 000[a]	12 458 000	2 771 000	15 229 000[a]
grand total (1969)				26 142 000	5 254 000	31 396 000

[a]Includes monies spent under Public Laws 81–875 and 84–99.

The complementary costs of slope damage are more difficult to estimate for two main reasons. First, such costs are reported by many separate agencies and a comprehensive picture is difficult to assemble; and second, damage cost estimates do not allow a clear distinction between floodway damage and slope-failure damage. In the City of Los Angeles, damage to private property was over $6 million, and probably ultimately exceeded $10 million (City of Los Angeles, Department of Building and Safety 1969, Slosson 1969), and most of this is attributable to slope failure (as the 'storm and slope failure damage reports' reveal). But most cost analyses simply generalise in terms of what was affected rather than how it was affected. For instance, the USDA Soil Conservation Service in the Antelope Valley (Cordell 1969) estimated crop damage and land restoration costs at about $6.5 million, but it is an assumption that most of this was caused by flooding. On a similar scale, the USDAFS (1969) reported that damage (repair costs) caused by both flooding and slope hazards in January in the Angeles National Forest amounted to over $6 million, of which nearly $5 million was required for road repair. On a smaller scale, the American Red Cross (Russell 1969) put the cost of disaster relief in January in Los Angeles and Orange counties at $20 000. The Red Cross activated 20 relief centres, assisted 2600 refugees, looked after 220 families after the crisis, and were helped by over 200 volunteers. Their help was given to anyone in distress, regardless of the cause of that distress.

The estimates of damage are therefore difficult to interpret in geomorphological terms. At the same time, they are difficult to interpret because the bases of assessment vary greatly: there are fundamental differences, for example, between an independent estimate of repair cost, a bid for relief based on a personal estimate, and a receipt for repair work; and the data are not comprehensive. Thus these cost figures give only a general impression of the damage caused by the 1969 storms. But undoubtedly a key feature of that impression (which contrasts sharply with the data for 1938) is the relative importance of slope-failure damage.

1978

Attitudes towards environmental problems changed substantially in the years following 1969, both nationally and locally. The National Environmental Policy Act (passed in 1969) initiated a period in which proposed federal, and later state and local developments had to be assessed in terms of their predicted impact on the environment. Locally, too, many minds were directed towards the dangers to the community of earthquakes, following the serious earthquake in San Fernando Valley on 9 February 1971 (Los Angeles County Earthquake Commission 1971). Such changes were of little direct consequence to the slope and sediment problems of Los Angeles County. Between 1969 and the storms of 1978, there was modest population growth from 7.0 million to 7.4 million (including a huge influx from Mexico), urban growth, and re-development. The trends continued towards the completion of flood- and debris-control plans and the refinement of slope engineering practices, but funds for such activities were limited. Thus the impact of the 1978 storms on the county was in many ways similar to that of the previous major storm episode in 1969, but somewhat less severe: there was relatively little flooding associated with the major controlled river systems, but debris clearance constituted a major expense; and slope failures were widespread.

Sources for studying the 1978 storms include reviews by the USACE (1978b), the LACFCD (1979a), and the National Research Council (1982), and detailed studies of slope failures by the California Division of Mines and Geology (1979), and the Association of Engineering Geologists (1978). Newspaper reports were consulted for the period, and Al Martinez of the *Los Angeles Times*, who wrote an excellent review of the storms (2 July 1978), also provided the summaries of storm damage data from both the City and County of Los Angeles. The storms were severe enough to establish a state of emergency in the City of Los Angeles (12 February), and state and federal declarations of a disaster area, although the problems were no more severe than in 1969. The following review is brief because of the similarities between the events of 1978 and 1969.

One of the most severe droughts in Californian history – in 1976–77 (State of California, Office of the Governor 1977a) – was broken by rainfall in December 1977; the rains extended intermittently into March, with particularly intense phases on 25–30 December, 9–10 February, and 3–4

DOI: 10.4324/9781003168645-9

March. The storms were well forecast, thanks in part to the new satellite technology (Garza & Peterson 1981). Only the last two of the three phases caused serious problems because the rainfall of the first phase fell on, and was largely absorbed by, the extremely dry ground. The cumulative rainfall curves for the period (Fig. 2.11) show that (as in 1969) precipitation generally increased with altitude, and that the two periods of maximum intensity (9–10 February, 3–4 March) both occurred after the thorough ground soaking in December–January and after the suggested 254-mm threshold of total antecedent rainfall for soil slips had been crossed. Without doubt, as the reports in newspapers confirm, most slope failures occurred in these two periods, or shortly thereafter. Locally, rainfall intensities were extremely high. For example, in Upper Haines Canyon, 35.6 mm fell in a 30-minute period on 9–10 February, and in the 3–4 March period many stations recorded rates of over 25.4 mm/h.

The total rainfall for the season had a recurrence interval of at least 20 years, and in places it exceeded 100 years. At older recording stations, 1978 ranked second to fourth, competing only with 1883–84, 1890–91, and 1940–41 (USACE 1978b). But runoff totals, at least in downstream locations, were not as great as in 1916, 1938, or 1969 – again a feature that was a reflection of the improved efficiency of flood-control measures.

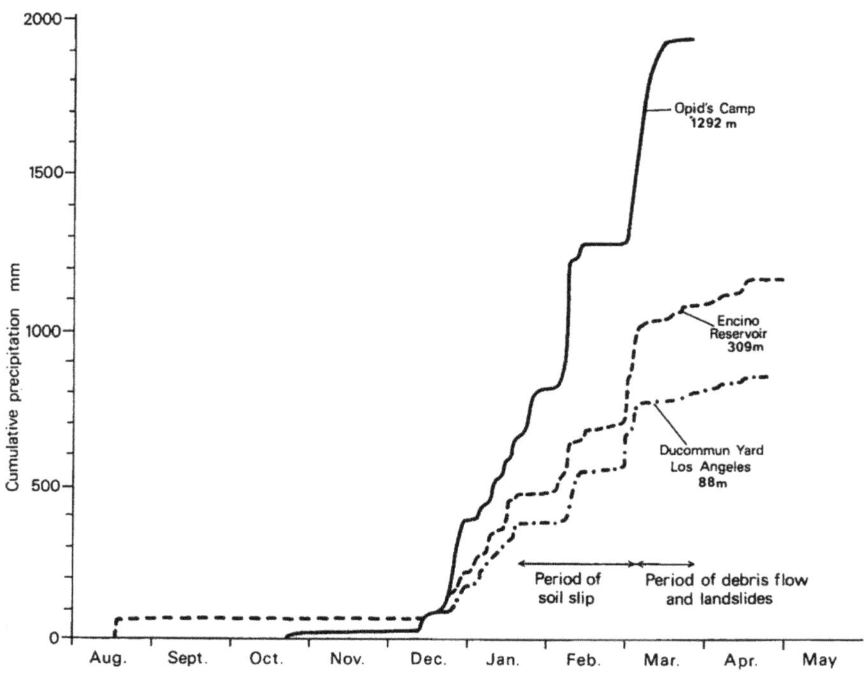

Figure 2.11 Cumulative precipitation curves for the storms of 1978 (after USACE 1978, J. Cobarrubias, personal communication 1979).

Runoff and sediment yield were so well controlled in the county (especially compared with adjacent counties) that area-inundation maps were not produced on this occasion. The only serious problems arose in headwater canyon areas of the San Gabriel Mountains, including Santa Clara Valley, Pacoima Wash, and the Big and Little Tujunga canyons. Here channel erosion caused extensive damage to houses, roads, parks, channels and flood-control works.

During the storms, some 920 000 m^3 of sediment were collected in debris basins (where yields ranged from 53 m^3/km^2 to 43 720 m^3/km^2, with an average value of 8 242 m^3/km^2), and about 6 700 000 m^3 accumulated in reservoirs behind dams (Davis 1982). The total volume of 7 600 000 m^3 was substantially less than the 14 388 363 m^3 accumulated in 1969, and the 15 527 052 m^3 accumulated in 1938.

Damage was most severe in the Big Tujunga drainage area and, as in 1934, in the small catchments on the north side of La Cañada Valley. The reason is clear. Over 20 000 ha had been burned in the region, especially in the Mill, Village and Middle Fork fires (1975–77), and the drought had retarded vegetation recovery. Thus the median sediment yield from the burned area was 26 240 m^3/km^2, whereas that from unburned areas was only 4220 m^3/km^2 (Davis 1982). The highest recorded yield of 43 720 m^3/km^2 was in Ward Debris Basin.

Reports mention mudflows in the valleys of the burned areas, where inadequate storage area, inadequate channels, or culvert blockage led to localised flooding. Pine Cone Road, La Crescenta, suffered damage from the overflow of Shields Canyon Debris Basin (which accumulated 34 600 m^3/km^2 during the storms). Zachau Canyon Watershed was completely burned by the Mill Fire, and heavy rainfall in the area led to a debris-laden flood that filled the Zachau Debris Basin, overflowed it, and caused flooding downstream in Sunland. Davis (1982) estimated that the total debris yield from the basin in 1978 was considerably greater than the recorded value of 41 860 m^3/km^2, and may have been as high as 59 000 m^3/km^2.

The greatest devastation occurred in the Big Tujunga Canyon, whose watershed had been partly burned by the Middle Fork fire in July 1977. Hidden Springs, a settlement poorly located in the floor of the canyon, was largely destroyed, and 10 people were killed by a debris flow on 9–10 February. The events were reminiscent of those of 1934, but the accounts of them are too imprecise for detailed comparisons. Certainly some houses were buried without disintegrating, suggesting debris flows (e.g. the *Foothill Ledger*, 11 February 1978). But the general impression is one of more fluid flows, discharging at the mouth of the canyon at about 255 m^3/s, and including a flood wave at Hidden Springs perhaps 4.5 m high (Davis 1982).

Undoubtedly these debris-laden floods would have caused more damage were it not for the emergency clean-out operations of the LACFCD that restored the debris-retention capacity of debris basins between the February

Table 2.11 Summary of flood damages in Los Angeles County, 1978 (USACE 1978b).

	Physical damages ($)	Business and emergency losses ($)	Total ($)
Property type			
highways, bridges and railroads	2 036 000	3 000	2 039 000
utilities			
sewer	34 000	31 000	65 000
water	83 000	25 000	108 000
other	—	—	—
public properties	312 000	115 000	427 000
flood control and channels and streams	31 179 000	719 000	31 898 000
totals	33 644 000	893 000	34 537 000
residential property damage[a]	60 200 000		
total	93 844 000		94 737 000
Drainage basin			
Los Angeles River	17 674 000	809 000	18 483 000
San Gabriel River	13 443 000	16 000	13 459 000
Santa Clara River	554 000	35 000	589 000
Santa Monica Basin	311 000	11 000	322 000
Antelope Valley	1 355 000	344	1 355 000
loss of life 16[b]			

[a]Based on reports by Los Angeles County and the City of Los Angeles for private property only.
[b]Estimate made by the author from newspaper reports.

and March storms, and the advice given to property owners following the fires on measures (such as sandbagging) they should undertake to protect themselves from flooding (LACFCD 1979a).

Table 2.11 summarises the costs of flood damage during the 1978 storms in Los Angeles County. The major element in flooding costs was that of removing debris from debris basins, reservoirs and flood-control basins. The USACE data suggest that $30.54 million of the total $34.53 million (88%) were for this purpose. It is important to emphasise these figures: for, while there may have been relatively little flooding and physical damage in 1978, the cost of controlling debris in Los Angeles County was (and still is) extremely high. The cost of the floods in Los Angeles County was relatively greater than elsewhere in southern California in 1978; damage in the whole region was estimated at $107 million. The USACE (1978b) estimated that damage *prevented* as a result of corps activities was $1200 million. Federal assistance to repair damage came principally from the Federal Disaster Assistance Administration, the USACE, the Federal Highway Administration, and the USDA Soil Conservation Service Emergency Watershed Protection Program.

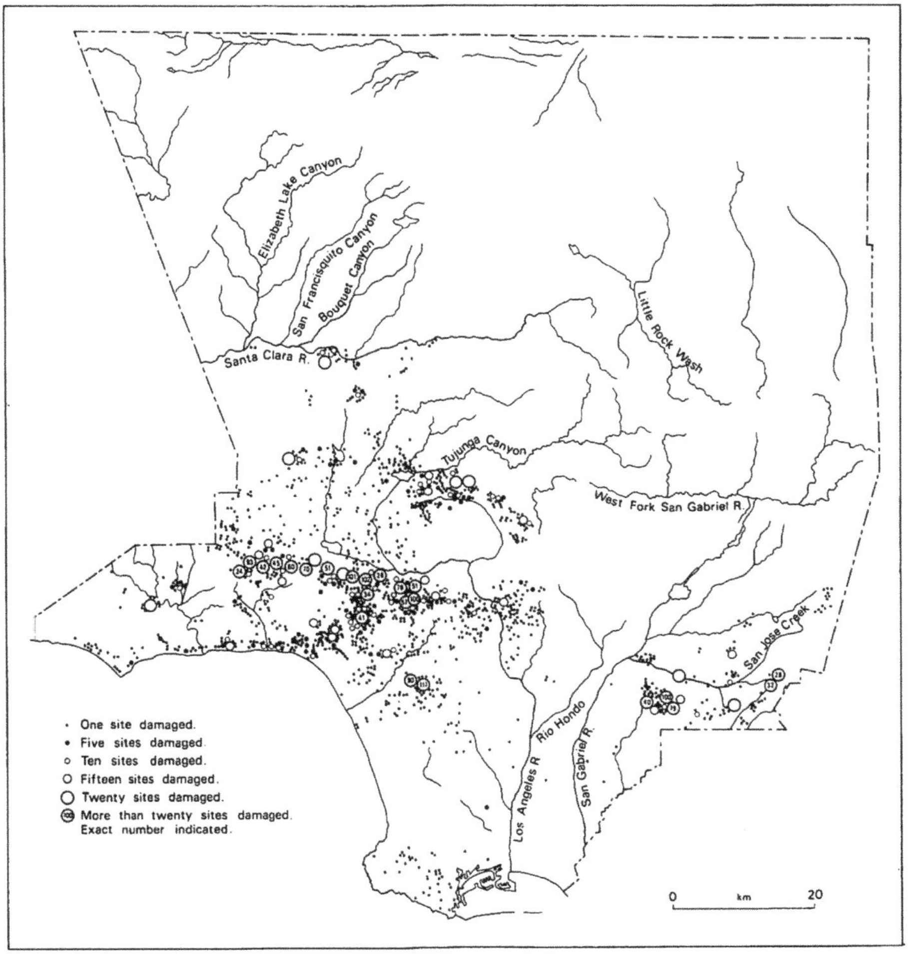

Figure 2.12 Location of storm and slope-failure damage reports for private properties collected by the building and safety departments of the City and County of Los Angeles, 1978. (Map by D. Clement, 1978. Copyright, 1978 *Los Angeles Times*. Reprinted by permission).

Most of the FDAA funds were used by the LACFCD to repair and clean out flood-protection facilities.

Slope damage was extremely serious in 1978, and constituted a higher proportion of estimated total damage than in 1969. Figure 2.12 (which is compiled from storm and slope failure damage reports for private properties collected by the building and safety departments in both the City and County of Los Angeles) gives an impression of the distribution of damage. (The data do not include natural failures in undeveloped areas.) Slope damage was, as in 1969, most serious in the Santa Monica Mountains, especially in the 3–4 March period, causing slope failures in communities such as Studio City, Sherman

Oaks, Encino, Tarzana, and Woodland Hills in the north, and Hollywood, Beverly Hills, Bel-Air, and Malibu in the south. There was also damage in Highland Park and the higher ground east of downtown Los Angeles, and in the Puente and San Pedro hills, as in 1969. In addition, however, considerable damage was reported in the new communities on the alluvial hills of the Los Angeles Coastal Plain, especially the Baldwin Hills, and on the flanks of the San Gabriel Mountains that had recently been burned. Extensive soil slippage was also described on Santa Catalina Island, where the intense rainfall attacked a surface made more vulnerable to failure by over 100 years of sheep grazing that had helped to transform large areas of coastal sage shrub and pine–oak woodland to bare ground or grass cover (Brumbaugh *et al.* 1982).

The California Division of Mines and Geology (1979) compiled a thorough review of 350 major slope failures during the 1978 storms, and analysed briefly the relations between failures and geology in the major hilly and mountain areas. For example (Fig. 2.13), they showed that the different types of damage in the Santa Monica Mountains overlap the major geological boundaries, and that failures were associated with the fracture zone of the Malibu fault zone and with block glides along the northward-dipping bedding planes in the Modelo Formation on the north flanks of the mountains.

The great majority of reported failures were (as in 1969) soil slips and associated debris flows that occurred on natural, 'cut' and 'fill' slopes. According to J. Cobarrubias (geologist in the City of Los Angeles Department of Building and Safety), these failures were mainly in the period 20 January to 4 March; this is the period which coincided with high rainfall intensities beyond the minimum rainfall threshold indicated by Campbell (1975); after 4 March debris flows and landslides were common, the latter having longer lag times than soil slips (see Fig. 2.11). A major rock slide closed the Pacific Coast Highway west of Topanga Canyon on 13 April (Forsyth & McCauley 1982).

The number of private addresses in the City of Los Angeles with slope-problem reports (as listed, for example, by the Department of Building and Safety on 31 December 1978) was about 3000. The county listed 1809 addresses. Loss valuation in the City of Los Angeles was estimated at *c.* $50 million; in the County it was $10.2 million. In their review of storm damage, the Association of Engineering Geologists (1978, see also Slosson & Krohn 1979) showed that of the failures in the City of Los Angeles, 93% (2790) were on sites developed before the 1963 grading ordinances, and only 7% (210) were on post-1963 graded sites. Furthermore, they showed that most damage to post-1963 sites was due to shallow slope failures, especially in engineered fills and/or natural slopes, whereas damage to older sites tended to be more severe and more expensive. Of the failures, 28% were soil slips and surface erosion, and 30% were 'mudflows' and 'debris flows' (geomorphologically, these were often associated with the downslope, more fluid portions of soil slips). Of the remainder, 22% were arcuate landslides or slumps on pre-1963 slopes or natural slopes, 8% were reactivated failures on pre-1963 sites, and 5% were

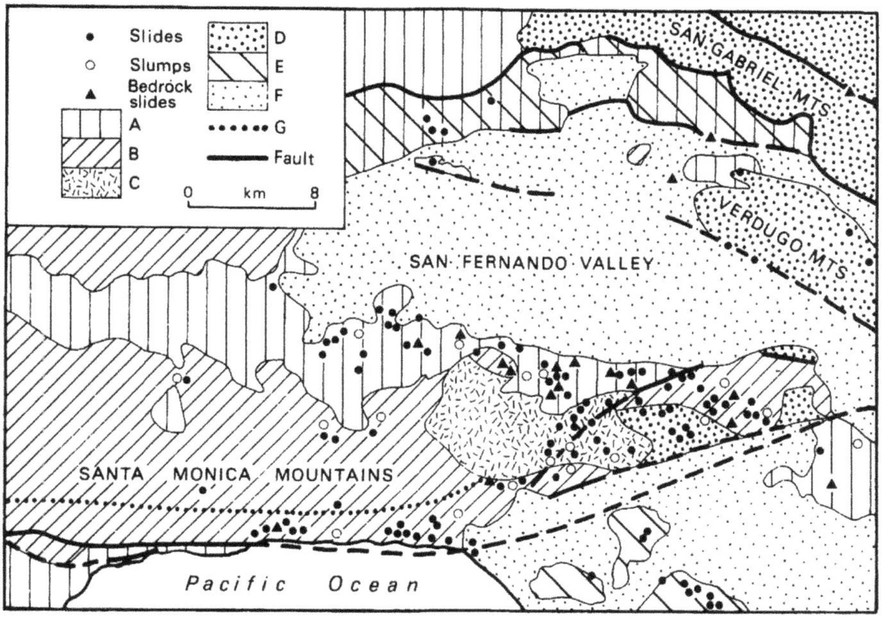

Figure 2.13 Location of major recorded slope failures in the Santa Monica Mountains in 1978 and related geological features. A, Modelo, Puente, Monterey and Fernando formations (most shales); B, Topanga, Sespe and Vaqueros formations (mostly sandstones), Upper Cretaceous and Paleocene rocks (mostly sandstones) and Conejo volcanics; C, Santa Monica Formation (slate); D, granite gneiss; E, poorly to moderately consolidated marine and non-marine silty sand and fine gravel; F, unconsolidated to poorly consolidated non-marine sand and gravel; G, approximate northern boundary of highly fractured belt of rocks along zone of Malibu Coast, Hollywood and Santa Monica faults (after California Division of Mines and Geology 1979).

new bedrock slides on pre-1963 sites. All deaths related to slope failure were associated with mudflows (Slosson & Krohn 1979).

The 2790 pre-1963 failures represented 7.5% of construction site permits issued, and were valued at $48.49 million; the 210 post-1963 failures represented 0.7% of permits issued, valued at $1.2 million (Slosson & Krohn 1979). Again, the progressive strengthening of building codes was vindicated in the city, and most damage could reasonably be attributed to inadequate earlier development. In the county the critical dates of code revision are 1962 and 1971. Of failure sites, 20.5% were developed after the modern building code revision in 1971 – a much higher figure than the post-1963 figure of 7% in the City of Los Angeles, and one that reflected the need for more effective inspection in the county (Slosson & Krohn 1979). Of the slopes graded between 1962 and 1971 in the county, 1.4% failed; of those graded since 1971, only 1.1% failed (Slosson & Krohn 1979).

While the patterns of slope failures in 1969 and 1978 are very similar, there is

Table 2.12 City of Los Angeles summary of storm damage and mitigation measures, 1978 (Association of Engineering Geologists 1978).

Storm damage classification	Occurrence, 1978 (%)	Problems	Solutions
I Mudslides (debris flows)	30	High probability for occupant injury; failure associated with ravines on natural slopes of 3 : 1 to 1 : 1	Develop techniques to recognise areas subject to failure type. Code revisions to subdivision rules for placement of drainage easements on side property lines with ravine outlet and concrete channel drains in the easement
II Soil slip-off	28	Universal occurrence related to shallow root development of ground cover and expansive-type soils	Re-investigate approved plant list to insure deep-rooted and low irrigation requirement. Homeowner maintenance information to be provided
III Shallow slump (arcuate) on natural and pre-1963 code fill slopes	22	Caused by uncontrolled drainage and/or loose fill at surface of fill slopes, predominantly at $1\frac{1}{2}$: 1 slope gradients	Develop drainage maintenance and inspection program by City Grading Inspectors. Correct uncontrolled upslope area drainage
IV Reactivated landslides	8	Primary occurrence on pre-code graded slopes and natural slopes. Original problem not fully solved in past remedial work	Repair slide by grading and correct drainage. Partial repair viewed only as a temporary improvement and not complete solution, as complete repair is not usually feasible
V Shallow slump (arcuate) on post-code slopes	7	Occurrence entirely on fill slopes exceeding 2 : 1 in gradient	Require new fill slopes predominantly at gradients of 2 : 1. Grading contractor to improve techniques in compacting slope surface. Limit new construction to setbacks at top and toe of slopes. Review methods used in slope stability evaluation by consultants
VI New landslides	5	These are existing graded areas with site conditions not conforming to current code standards. Problems amplified by existing construction at top and toe of slopes	No new construction before repairs are completed. Establish City–County inspection of these areas for drainage and related problems. Public information to provide criteria for buyers of hillside property

a shift of emphasis in the reports and subsequent discussions. In 1969, it can be argued that the soil slip was clearly seen as the principal villain, whereas landslides had seemed more important previously. In 1978, debris flows on slopes were perceived as a more serious problem than in 1969 (e.g. State of California, Office of the Governor 1978). In part, as Slosson and Krohn (1982) suggested, this change may have reflected a real increase in debris flows as a result of recent building development at the base of slopes associated with older subdivisions, and inadvertent but approved construction on natural slopes subject to debris flow hazard. In part, too, it may have reflected the fact that while such flows had occurred previously, they had not been recognised or studied extensively, whereas other specific features (such as soil slips) had. The aftermath of the storms included the establishment of professional review committees in both Los Angeles County and the City of Los Angeles (e.g. Slosson & Krohn 1979). These committees made many recommendations for improving responses, especially in terms of further strengthening development control through increased site inspection, and in terms of specific proposals designed to solve problems arising from specific types of slope failure. Table 2.12 illustrates the nature of the solutions proposed for the City of Los Angeles (Association of Engineering Geologists, 1978).

CONCLUSION: BEFORE AND BETWEEN CRISES

The storms, floods and slope failures of 1914, 1934, 1938, 1969 and 1978 were the most important in the 20th-century history of Los Angeles County in terms of their impact on the communities and their effectiveness in generating management responses. But there have been other geomorphological events, some of which signalled important management changes and improved public perception of geomorphological processes.

The history of floods before 1914 is not well documented (Los Angeles County Board of Supervisors 1915, Troxell *et al.* 1942), but the combined testimony in the case of *Daneri v. The Southern California Railway Company* in the Superior Court of Los Angeles (1897), and the reports by Reagan (in Los Angeles County Board of Supervisors 1915) and Eaton (1931) record flooding in the Los Angeles and San Gabriel rivers in 1811, 1815, 1822, 1825, 1833, 1842, 1851–52, 1859–60, 1862, 1867, 1884, 1886, 1887, 1889, 1890, 1891, 1905, 1906, 1909 and 1911 – a frequency of approximately once in every five years. Of these floods, those of 1884 and 1889 are generally reckoned to have been the most severe.

In the early years, flooding affected huge areas in the Los Angeles coastal lowlands, which were, in effect, a series of enormous alluvial fans across which precise flow directions were relatively unpredictable – as they were on the smaller alluvial fans near the mountains later in the 20th century. Thus, the Los Angeles River changed directions during floods on several occasions. In 1815 it moved from an easterly course into a more westerly one, draining into Ballona Creek and Santa Monica Bay. In 1824–25 it broke eastwards again to discharge into San Pedro Bay. In the great floods of 1861–62, the San Gabriel River also shifted course, from the east to the west of El Monte. And in the floods of 1867–68, the San Gabriel River split above Whittier Narrows to create a new channel to the sea at Alamitos Bay; the old channel, to the west in San Pedro Bay, became the outlet of the Los Angeles River, although it still received water and sediment from the San Gabriel system via Rio Hondo. The double storms of 1884 led to two floods (a phenomenon familiar to the 20th century), causing serious damage to the now rapidly growing city. And the floods of 1889 were probably as severe.

The year of 1914 was certainly an administrative watershed, but the major floods of that year were followed by only local minor floods in 1916, 1918, 1921–22, 1926, 1927, 1931, 1932, 1936 and 1937 (Troxell *et al.* 1942) before the

DOI: 10.4324/9781003168645-10

major regional floods of 1938. The 1934 floods were also local, but with major regional implications.

Between 1938 and 1969, the balance of emphasis shifted from major flood problems to major slope-failure problems. There were, it is true, some very wet winters and some serious floods. The 1940s included some exceptionally heavy storms (especially 1940–41 and 1943–44), but because they were preceded by relatively little rain, their consequent discharges produced little flooding.

In 1952 public attention first began seriously to focus on slope failure. Although the flooded area during that especially wet season amounted to about 6500 ha, and sediment problems were associated with burned areas in San Fernando Valley (such as Brown's Canyon, burned in 1950 – USDA 1953), landslides (especially above Hollywood) were the most serious problem (LACFCD 1952). Thus in 1952, and again in 1962 and 1965, as in 1969 and 1978, damage to slopes and to urban developments on slopes accompanied the seasonal storms.

Not all slope failures were contemporaneous with, or even caused by, the major seasonal storms. Indeed, the early perception of slope failure problems in Los Angeles County was perhaps as much conditioned by a small number of spectacular landslides as by the more numerous, but smaller, winter storm failures. In the context of larger landslides, two areas are notorious.

The first is the Pacific Palisades, an area of steep cliffs and canyon bluffs in poorly consolidated Cretaceous–Quaternary clastic sediments north of Santa Monica. The Pacific Coast Highway, carved into the base of the cliffs (and the most important recreational artery in the metropolis), has been plagued by substantial and serious slope failures for many years (McGill 1959, Cleaves 1969, USACE 1976). There are over 50 active landslides, together with historic and pre-historic slides throughout this desirable residential area (McGill 1973). The earliest record of a slide is in 1874. A detailed 20th-century record of landslide movements was compiled by the USACE (1976). It shows that intermittent slope failure in the area is long established, and that particularly important dates are 1938, 1940–41, 1948, 1952, 1956–60, 1962 and 1969. It also shows that the frequency of failure reports increased in the 1920s and 1930s, subsequently reaching a 'plateau with peaks' (Fig. 2.14): this may truly reflect an increase in landslide movement, or it may reveal an improvement in reporting. (For comparison, the frequency of landslide reports in the *Los Angeles Times* is plotted on Figure 2.14. This record also shows an increasing frequency since 1937, but the trend is only loosely related to landslide frequency reports from the Pacific Palisades.) Of the numerous events at the Palisades, the most traumatic were those in 1956–58, especially the Via de Los Olas (or Flagg's Restaurant) slide which moved 589 680 m^3 and blocked the Pacific Coast Highway in 1958, and the slides of 1965 which caused $1 019 000 damage (USACE, 1976).

The second notorious area of slope failure is along the terraced seaward

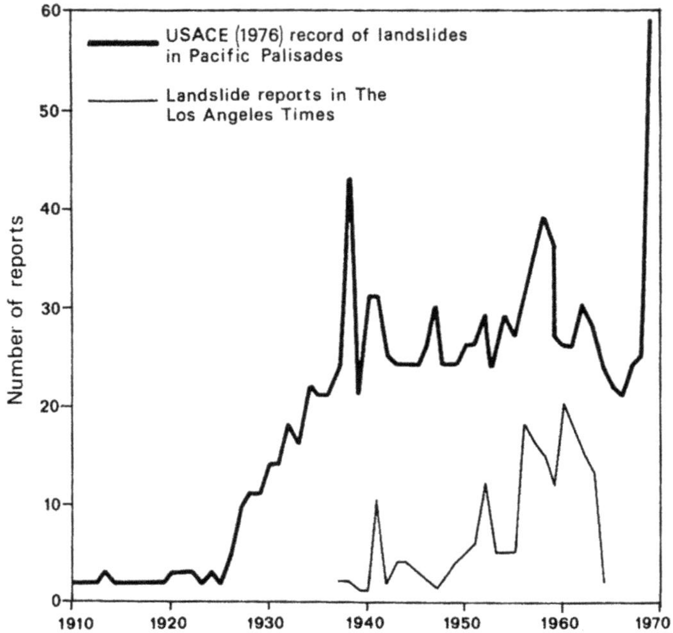

Figure 2.14 Frequency of landslide reports in the Pacific Palisades (based on data in USACE 1976) and in the *Los Angeles Times* (data source: the *Los Angeles Times* library).

slope of the Palos Verdes Hills. There are many failures here, both ancient and modern (e.g. Jahns & Linden 1973, Ehlig & Ehlert 1978). The most serious failures are at Portuguese Bend, where there is essentially an enormous complex block glide (Pipkin & Ploessel, no date) comprising a typical suite of landforms that includes hummocky terrain, arcuate scarps, poorly drained depressions, and extensive surface cracking (Merriam 1960). The feature had been active in the Pleistocene, and was known to exist before urban development began. It was initially caused by marine erosion at the base of the cliffs that affected a seaward-plunging synclinal structure in rocks dominated by the Miocene Monterey Formation, and in particular by tuffaceous Altamira shales within it (Merriam 1960, Jahns & Linden 1973). But the slide was relatively inactive until 1956, when intermittent movement began. The movement extended in area and varied in rate, with peak rates apparently lagging behind winter precipitation peaks by a month or two. It seems probable that the movement was initiated by winter precipitation cruxially complemented by an influx of water from houses built in the area a few years before 1956 (Merriam 1960), and promoted by a variety of factors including earthquakes, high tides and increased load caused by superposition of fill. The movement caused serious damage (in excess of $10 million) to over 150 properties and the major road through the area (Easton 1973), and led to headline-catching law suits.

The saga continues of slope failure and flooding accompanying storms, and landslide activity following them. In 1980 the City of Los Angeles Department of Building and Safety received at least 3397 calls that required site inspections following the rains of January and February (U.A. Foth, personal communication, 30 April 1980). Again, the pages of the *Los Angeles Times* reverberated to pictures of personal tragedies in Baldwin Hills, Sherman Oaks, Mandeville and Topanga canyons and other hazardous locations. Over 13 people were killed, 34 homes were destroyed, and hundreds of threatened residents were evacuated.

Such crises – some short and sharp, others more prolonged – provide the essential stimulus for improved community and management responses.

3
CONSEQUENCES

Response and responsibility

Conflicts

When a natural disaster strikes a settlement, the communities' responses are a complex reflection of the physical nature of the hazard, and the varied ways in which individuals, interest groups, and the whole population perceive the problems and their solutions. Individuals are influenced by the extent to which they are directly affected (as sufferers, innocent bystanders, or as managers, for example), by the nature of the events themselves (such as their magnitude, intensity, and duration), and by their memories, experience of the problems, and social characteristics (education, age and sex, security, and family commitments etc.). In general, the individual affected by the hazard seeks peace of mind, or to minimise loss, and personal responses reflect these ambitions. Crudely put, the individual may choose to 'fight' or 'take flight' – to resist the hazard, or either to move away from it or to transfer the responsibility for suffering and its alleviation to a higher authority (such as the government). If the individual fights, a range of actions may be open and the weapons may be sufficient to restore peace of mind or reduce losses to an acceptable level, regardless of whether the problems are solved. If an individual's faith is placed in higher authority, he or she may also have peace of mind restored or losses alleviated. But if both individual and community efforts fail to solve the problem, the individual will probably respond by further adjustments or by accepting only a lower level of satisfaction.

As with individuals, so with groups in the community. In addition to the individuals who suffer directly from storm damage in Los Angeles, there are numerous interested groups. These include homeowners' associations, city, state and federal agencies with responsibilities for certain aspects of the problems, consultancy and research organisations, pressure groups, and those commercial interests involved in land, its sale, development and protection. Differences of perception within this array of groups stimulate numerous conflicts. The following discussion briefly illustrates the nature of these conflicts.

Apportioning blame often promotes conflict and litigation. Few individuals affected by storms blame themselves for their misfortune. Some believe they are adequately protected. For instance, a Cloverdale woman put her faith in the strength of ivy: 'You know, we had ivy on that hillside. I thought just Friday we

DOI: 10.4324/9781003168645-11

were protected. But I guess it was undermined . . .' (*Los Angeles Times*, 5 March 1978). Others seek human scapegoats. Some may claim that the builders or the local authorities were negligent. Management agencies may plead shortage of time, money and personnel, blame homeowners for poor property maintenance, and emphasise the need for research. And developers may accuse subcontractors, homeowners and local authorities.

Then there are conflicts of political philosophy and personality. For example, during the crises of 1969, the philosophical conflict between 'self-help' and 'public assistance' was brought sharply into focus by Governor Reagan – who was attempting to prevent an increase in state gasolene tax to raise funds for flood relief, and was asserting his view that people should attempt to solve their own problems; and by Mayor Bradley – who was advocating the use of funds external to Los Angeles for disaster relief. In touring the disaster areas, Governor Reagan declared:

> Once upon a time, when a barn burned, we used to have a barn-building party and erect it . . . I don't think there's anything wrong with a little neighborliness. I'd like to see some flood relief fund-raising drives get going (*Los Angeles Times*, 29 January 1969).

Mayor Bradley retorted that things had changed since those times, and some of the storm problems were too big for such self-help solutions.

Closely related to political philosophy is the conflict over who should pay to solve problems. Is a local hazard, a personal, local authority, county, state or federal responsibility, or is it the responsibility of one and all? Intimately associated with this question is the matter of how much should be made available for assistance. The case of the infamous 'Proposition 13' well illustrates this problem. Proposition 13 advocated dramatic reductions in property taxes; after much debate, it was put to the electorate in July, 1978, and accepted. The consequences for hazard management were immediate. For example, the LACFCD had its revenue reduced, and was obliged to cut its staff and slow down its debris-basin clear-out programmes. Perhaps unwittingly, the electorate had reduced its hazard protection. Another aspect of the conflict over funds is the rivalry for them that may arise between hazard management agencies.

Conflicts arise between the demands for responses to geomorphological hazards, and the demands for responses to other significant hazards and problems facing the community. It is the rôle of the political process to strike a balance between these conflicting demands. The broader context of environmental hazards requires emphasis here for it has strong spatial and temporal components. In a series of important studies, Van Arsdol and others (1964, 1975) and Van Arsdol (1969, personal communication 1975) showed that the 'artificial' environmental hazards of smog and air-traffic noise in metropolitan Los Angeles were very widespread, affecting some 50% and

25% respectively of the population in the 1960s, and particularly affecting older residential districts, which also tended to be somewhat poorer and burdened with serious social problems. In contrast, the 'natural' hazards of fire, slope failure and flood affected approximately only 2%, 15% and 16% of the population respectively (with considerable spatial overlap), and tended to influence younger, more affluent and 'articulate' communities. The spatial and temporal variability of environmental hazards has important political implications because the constituencies of pressure groups constantly reflect it. In addition, while the political response *tends* to be proportional to the extent to which a hazard is *perceived* to be serious, perceptions vary. For example, a severe sporadic event (like a flood), associated as it often is with institutions designed to control it, personalised warning information and the immediacy of danger, is seen to be relatively more important than other, more continuous problems. Similarly, a vociferous, affluent, well-educated and influential minority (from the fire–slope-failure area of affluent Malibu, for example), because it is more likely to complain and to be listened to, may elevate a problem to a level of political importance beyond that which it deserves. And, without doubt, environmental hazards at times pale to insignificance when set against such fundamental social hazards as divorce, crime on city streets, and unemployment.

Conflicts also erupt over the interpretation of problems and over the adoption of solutions. Such conflicts are commonly expressed through political lobbies and pressure groups whose motivation ranges from altruism to greed. Almost every proposed solution carries the seeds of conflict, and with good reason, because not only is one man's solution often to the detriment of another, but there are commonly differences of opinion about the effectiveness of proposed solutions amongst professional environmental managers – largely because of incomplete knowledge of the highly complex problems. One example (of many) concerns efforts by councilmen in the City of Los Angeles to introduce a slope-density ordinance that would eliminate construction on certain steep slopes and limit density on others, thus reducing the potential for slope-failure damage (City of Los Angeles 1975). The response of the building industry was unequivocal (Martinez 1978). One New York developer sued the City of Los Angeles in order to invalidate such restrictions already introduced in two communities in the Santa Monica Mountains, on the grounds that they violated the city charter and were discriminatory because they applied to only one area of the city. Building Industry Association lobbyist Richard Worth declared: 'It [the 1978 storm] was a hundred-year rain. We didn't cause it . . . Things failed because homeowners have tampered with the grade . . .'. Instead of new restrictions, 'they ought to analyze the areas that failed, put together a fund and have this industry to grade and buttress it . . . but we'll have nothing to do with slope-density because its a sham' (quoted in Martinez 1978). Density may, of course, be positively correlated with profit.

Conflicts also arise between short-term and long-term solutions. The former are usually more politically acceptable than the latter, perhaps because immediate crisis responses demonstrate the concern, dynamism and responsibility of legislators more effectively than long-term and less visible solutions. Similarly, disagreements between different administrative agencies also appear from time to time, often as a consequence of their separate responsibilities and experience. An example arose in the 1950s and early 1960s within the City of Los Angeles between the Department of Building and Safety (which was charged with responsibility for private property) and the Department of Public Works (which was responsible for city property such as roads and public buildings) – Jahns (1969). Both departments were innovators in slope control, but their different responsibilities led to the development of different philosophies, regulations and procedures, and to an undesirable degree of duplication.

Patterns of response

The complex political and managerial procedures for evaluating hazards, and identifying, choosing and implementing solutions, are simplified in Figure 3.1. The evaluation of a hazard event arises normally from a request for politically acceptable answers to the questions that are asked immediately following the event: 'What happened, and why, and how can we best prevent the problems recurring?' The answers commonly come from post-mortem advisory committees that make recommendations and provide the basis for further

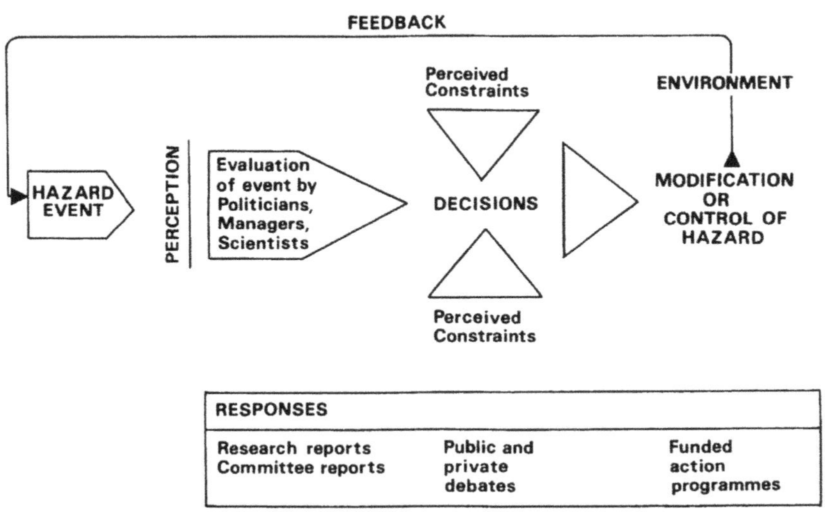

Figure 3.1 Model of response to a hazardous event. (Developed from Kasperson's general model of municipal stress management, Kasperson 1969).

research. Public and private debate follows, usually culminating in decisions, legislation and action programmes.

Public decisions about action in response to environmental stress can take a variety of forms. Table 3.1 provides a classification based largely on the ideas of several authors working elsewhere (Sorensen *et al.* 1975, Burton *et al.* 1978, Ward 1978, Mitchell 1979, Smith & Tobin 1979) and on local examples. In many ways, hazards may be viewed as negative resources carrying a cost both to individuals and groups. Therefore, a major class of response involves paying the hazard's price: by the individual bearing the loss, by sharing the loss amongst the community, or by planning for the loss in advance through insurance policies. Attempts to reduce losses by improved understanding are based largely on scientific, engineering, and social science research, and

Table 3.1 A classification of adjustments to geomorphological hazards in Los Angeles County.

	Examples	
Class of adjustment	Slope problems	Flood/sediment problems
A. Paying the price		
1. Bear the loss	Individual loss bearing	
2. Share/spread the loss	Public relief and rehabilitation (state or federal funds)	
	Tax adjustments	Tax adjustments
	Litigation	Litigation
3. Plan for loss	Landslide insurance	Flood insurance
B. Improve understanding to reduce losses		
4. Evaluation, prediction, monitoring, forecasting	Monitoring slope stability Mapping slope stability characteristics	Analysis of historic sediment yields, rainfall-runoff relations
C. Modify the hazard to reduce losses		
5. Emergency responses	Slope drainage	Debris clearance
	Evacuation	Sand-bagging
6. Improve preparations	Emergency procedures	Emergency procedures
7. Modify the cause		
Prevention	Limit time of slope construction	Abatement: watershed vegetation control
Correction	Improve drainage	Revegetate burned areas
	Buttress slopes	
8. Control the effects		
Planning	Building codes	Floodplain development control
	Zoning and subdivision ordinances	
Engineering	Improve foundation design	Debris basins
		Dams
		Flood-control reservoirs
		Channel improvements

monitoring and evaluation figure prominently here in attempts to predict and forecast events, and to gauge individual and community responses. Losses may be further reduced by modifications to the hazard itself: through emergency action and improved emergency preparations, by modifying the causes of the hazards and by controlling their effects with planning controls or engineering works.

Legislation incorporating these responses may range from city ordinances to federal laws. The legislative acts are the tangible results of the conflicts in the political and managerial arenas, and often they reflect a compromise between development interests concerned to maximise profits and opportunities, and the community exercising what it presumes to be its social responsibility. The compromises vary from place to place and within the political hierarchy, and they vary too, from legal requirements placed on individuals to formidable investment plans.

A management hierarchy

Political decisions are mainly implemented through management agencies. There is a hierarchy of such agencies in Los Angeles County, ranging from those in individual cities, through county and state agencies, to federal agencies. In addition, there are several private management groups. In the context of slope, flooding and sediment hazards, the agencies have different responsibilities, which can be classified into, first, emergency action, second, survey, planning, assessment and insurance, and, third, repair, maintenance and construction. Figure 3.2 represents an attempt to summarise some of the relationships between these responsibilities and the major hazard management agencies in Los Angeles County. The rôles of each stratum in the management hierarchy require brief review.

PRIVATE AGENCIES

Amongst private groups concerned with geomorphological problems, consultant engineering geologists figure prominently in the analysis of site conditions prior to and during development and of site failures after storms. Some consultants (especially those with academic affiliations) contribute through their site-specific experience to a broader understanding of the regional problems. Their work is partly fortified by statutory requirements for geological site evaluations, which are now embodied in many municipal building codes. The consultant's rôles are varied, ranging from technical advisor or assessor, through field detective, surveyor and office researcher, to planning aide and legal witness (e.g. Leighton 1971). Flooding does not have a similar coterie of consultants, although sediment-yield problems and debris-flow hazards certainly require the attention of geologists. In southern

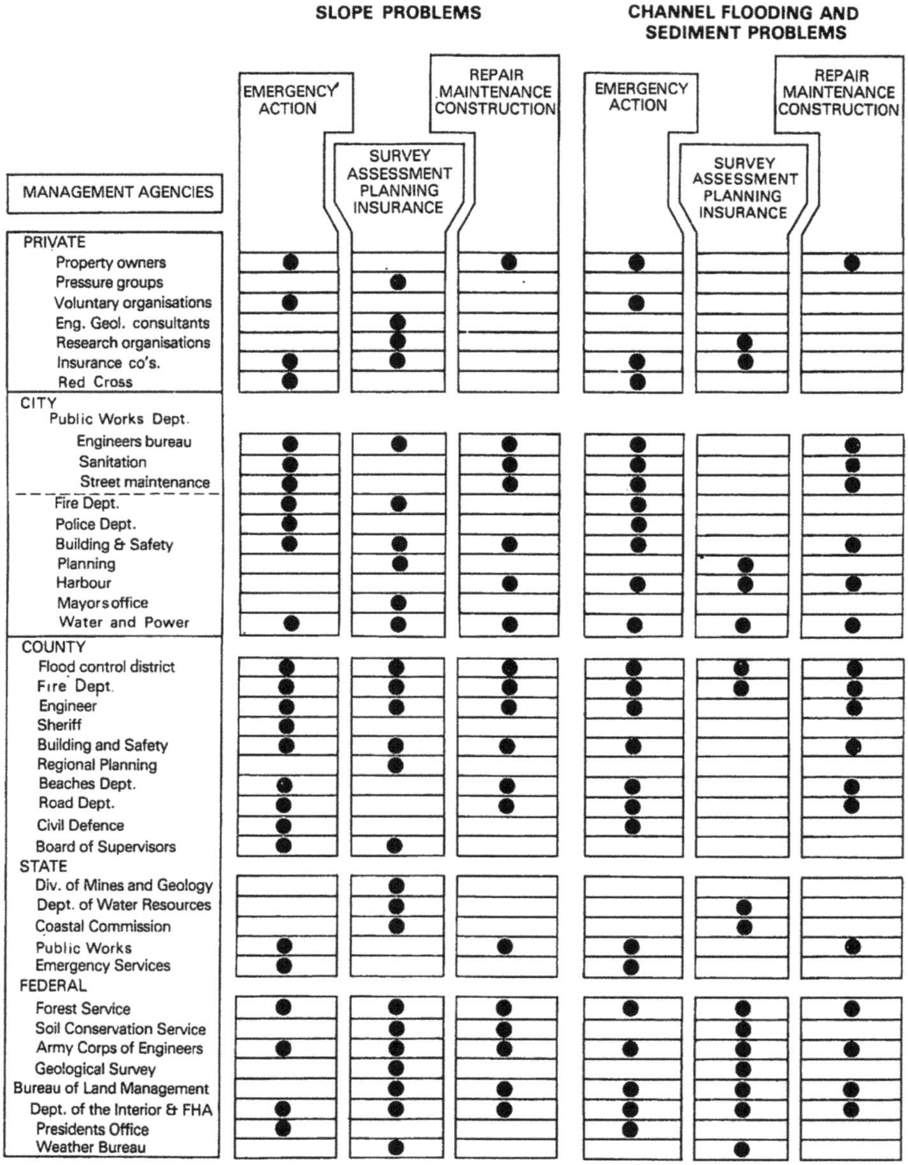

Figure 3.2 Some relations between the hierarchy of management agencies and slope, channel-flooding and sediment problems in Los Angeles County.

California, geological consultants come together (with others) in the significant pressure group of the Association of Engineering Geologists.

The rôles of other private groups interested in geomorphological hazards are more diffuse. Research organisations (such as universities) often provide independent assessment of problems (Merriam 1960, Pipkin & Ploessel, no date), and at times academic research developments can give a powerful impetus to the understanding of management problems. For example, the slope analyses of Strahler (e.g. 1950) in the Verdugo and San Rafael hills and elsewhere, were seminal to the development of sediment-yield predictions (e.g. Lustig 1965). Another important group of private interests is found in residents' associations. These abound, and, not surprisingly, are particularly well organised and vocal in their self-defence in areas that have both slope and fire hazards and a well educated, affluent population. The Bel-Air Association, in the Santa Monica Mountains, is one example; the Miramar Homeowners' Association Inc. and the Castellamare Mesa Home Owners Inc. are two groups concerned with landsliding in the Pacific Palisades (USACE 1976).

CITY AND COUNTY AGENCIES

At the level of city administration, concern is primarily with control and development of sites, the repair and maintenance of facilities, and emergency responses to crises. There are now 81 cities in Los Angeles County, and each has its own structure and political philosophy, with the result that there is substantial spatial variation within the county in the ways that geomorphological problems are dealt with. The City of Los Angeles is the largest and amongst the oldest, and it is certainly in the forefront both locally and nationally in developing procedures to manage environmental hazards. In most cities, hazard survey, assessment and planning lie mainly in planning departments and the building and safety (private property) and public works (public property) departments. The latter agencies also respond during and after emergencies. And the police and fire departments usually carry a heavy responsibility for public assistance during the emergencies themselves. Geomorphological advice within city agencies comes mainly from the growing number of engineering geologists, who make comments on consultant reports, on proposals for tracts, grading and construction, and on geological aspects of litigation and planning development; they also consult with developers and contribute towards the improvement of building and planning regulations (Yelverton 1971).

Los Angeles County is concerned with geomorphological problems at two levels. First, it acts as a city authority for unincorporated county land and it provides certain services for those cities within the county that so choose. For example, in 1977, 33 cities had an agreement with the county for building inspection, and 40 cities participated in the 'Consolidated Fire District'. Second, by agreement, the county provides a service to most areas through the LACFCD, one of the oldest organisations of its kind in the country. This agency is responsible for planning, constructing and operating many flood- and sediment-control facilities, for emergency management during storms, and for assisting with the planning of flood-zone development and insurance (e.g. LACFCD, no date). It includes engineering geologists on its staff.

STATE AND FEDERAL AGENCIES

If cities and counties are inevitably concerned mainly with local day-to-day problems, then the state and federal agencies are more concerned with longer-term perspectives involving survey, assessment and planning in regional contexts. In general, state and federal agencies work on their own terrain (e.g. the USDAFS, the Bureau of Land Management, and the armed forces are responsible for large tracts of land in Los Angeles County), on contracts tied to other agencies, or on specific issues of state or federal concern. For example, the geological services of the state's Division of Mines and Geology and the federal USGS have played a fundamental rôle in the regional appraisal of the problem of slope failure (Blanc & Cleveland 1968, Morton & Streitz 1969, Campbell 1973, Radbruch & Crowther 1973). An outstanding example of regional planning is the Urban Geology Master Plan for California (California Division of Mines and Geology 1973). There are clearly possibilities of conflict and competition between state and federal agencies in regional work, but these are reduced and the prospects of collaboration improved by exchange of current and proposed project and assignment lists, and, in the case of geological studies, by the Department of the County Engineer which administers a clearing-house of information.

State and federal agencies also fulfil important educational rôles. One of these is in bringing to the attention of hard-pressed local authority employees the fruits of others' experience. The use of demonstration sites by the California Division of Mines and Geology (Cleveland 1971b) is one example. Another is seen in the rôle of bringing together at conferences those with common interests in geomorphological problems from the whole range of agencies within the administrative hierarchy (e.g. State of California, The Resources Agency 1965, Office of Emergency Preparedness 1969, US Department of Housing and Urban Development/US Department of the Interior 1971). Regional professional organisations also serve this function (e.g. Association of Engineering Geologists 1966). In addition to these larger-scale, longer-term perspectives, both state and federal agencies have responsibilities for emergencies, especially as they affect their own property and, through the provision of funds.

Management responses

Emergency responses

EARLY WARNING

Early warning of a storm is based primarily on weather forecasts from the US Weather Bureau. Such warning may come directly or indirectly to those agencies responsible for storm management. In general, storm forecasting has improved over time, especially in recent years with the advent of weather satellites. Flood forecasting has also improved especially with the introduction of 'real-time flood warning systems' based on radio transmissions to local centres of precipitation data and computer-based predictions of peak stream flows (Bartfield & Taylor 1981). If the forecasts are made early enough, agencies and private citizens are able to take precautionary measures and to activate emergency programmes. Whether such responses are adequate is another matter.

Experience of storms shows that in addition to weather and flood forecasts there are two other important early warning signs: dry-season fires and early wet-season rains. The former locate areas where, if storms do occur, their impact is likely to be greatest. Residents in these areas can be alerted, for example by the LACFCD (Los Angeles County Fire Department 1969, LACFCD 1979a), and emergency protection (usually debris control works or reseeding of vegetation) may be initiated by agencies, or by residents possibly with assistance from groups such as the California Conservation Corps. Early wet-season rains often mean that soil-moisture storage is limited if and when heavy storms occur in January–March, so that the proportion of runoff generated by the storms may be enhanced. Such a sign may simply alert storm-management agencies; or it may encourage preparations such as emergency debris clearance from debris basins.

INTEGRATED RESPONSE

Geomorphological problems figure prominently in the range of urban emergencies that require rapid and complex institutional responses in Los Angeles County; other such emergencies include fire, earthquakes and,

DOI: 10.4324/9781003168645-12

potentially, wars. Fundamentally, two types of response are required to storm-generated hazards. The first involves dealing with people, damage and disruption; the second concerns the management of water and sediment discharge and of slope failures. Formal responses depend to a considerable extent on the perceived magnitude of the hazard, for there are perceived thresholds of magnitude which, when crossed, generate higher levels of response. As a result of many years of experience in Los Angeles County, and given the benefit of state and federal advice (State of California, Office of Emergency Services 1974, US Department of Commerce 1977), the region has now developed, codified and tested a coordinated series of responses which are normally effective.

The development of integrated emergency responses is associated with a substantial history of legislation. For example, federal acts include PL 81–875 and PL 81–920 (National Plan for Emergency Preparedness), PL 93–288 (Disaster Relief Act of 1974), and the President's executive order of 20 July 1979 created the Federal Emergency Management Agency. State legislation includes the California Disaster Act, the California Civil Defense and Disaster Plan, and the State of California Emergency Plan which incorporates the California Emergency Services Act. In addition, there are county and city regulations.

Emergency plans and services have been formally coordinated within the Los Angeles County Operational Area organisation since 1965–66. This organisation represents an integration of federal, state, county, city, special district and other agencies (including the American Red Cross and the Salvation Army), and it operates within the terms of the *Los Angeles County and cities disaster relief manual* (1970). (This and other manuals mentioned in this section are frequently modified. The dates given are those on the manuals consulted.) The rôles of some of the agencies that may become involved in county-wide disaster relief are summarised in Table 3.2. In addition, there are provisions for mutual aid on a regional basis. Los Angeles County is within the state's mutual aid region I, which also includes San Luis Obispo, Santa Barbara, Ventura and Orange counties.

The critical thresholds in the scale of perceived hazard magnitude – as defined in County Ordinance 10493 (28 April 1972) – are recognised as local emergency and state of emergency: the former identified and proclaimed by county (or, in some cases, city) authorities, the latter requested by the same authorities of the state governor (through the Office of Emergency Services) who may decide to make a proclamation. At the federal level, the president, through the Federal Emergency Management Agency (1979), may declare a 'major disaster' or 'emergency' on receipt of advice from the state governor, which in turn activates additional assistance. It should be emphasised that these thresholds, while restricted to particular types of event, are not scientifically specified. They are, in effect, based on initially professional, but subsequently political judgements. The various levels in the hierarchy of emergencies have all been activated in recent years. For example, all thresholds were crossed, and all systems used, in the storms of 1969 and again in 1978; in 1970, the fires in the mountains led to a proclamation by the governor of a 'state of emergency' in Los Angeles County and elsewhere.

Most individual organisations within the administrative hierarchy commonly have developed their own operational manuals. Thus, for example, the USACE (1978a) has a *Natural disaster activities* manual, the LACFCD (1964) has a *Storm operation guide* and several other manuals, and the Los Angeles County Fire Department (1967a) has an *Emergency operations procedure*. Independent groups (such as the Red Cross) also usually have clear emergency procedures (Russell 1969). Autonomous cities which do not contract emergency services from the county have their own procedures codified. For example, the City of Los Angeles Department of Public Works (no date) has an *Emergency and disaster manual* that controls the city response to damage on public and, in certain circumstances, private property.

PEOPLE, DAMAGE AND DISRUPTION

Emergency reactions to the ravages of storms arise partly from individual initiative and partly from institutional responsibilities. Individuals respond to personal stress in a multiplicity of ways, but in general they seek to fight the problem or to distance themselves from it. They may, for instance, adopt the 'spirit of the sandbag', taking measures with their friends and neighbours to protect their property; or they may flee, perhaps leaving others to fight for them. The pages of local newspapers focus on such responses and their related hardships and heroism, and on the vitality of self-help and improvisation. There are numerous instances of *ad hoc* organisations being created in crises. For example, an 'emergency task force' was formed by volunteers in Coldwater Canyon (Santa Monica Mountains) in March, 1978 (*Los Angeles Times*, 6 March 1978). The press and other media also play an important rôle in alerting the public and in advising them on procedures to follow. Public agencies also offer advice. For example, the LACFCD advises property owners in areas vulnerable to severe flooding following brush fires on measures they should adopt (LACFCD 1979a) and the Defense Civil Preparedness Agency (1976) offers more general guidance. Nevertheless, the rôle of self-help, despite its importance, is one that has not been fully appraised or effectively incorporated into institutional plans.

Institutional emergency responses to the needs of individuals and their property mainly involve the police, fire and hospital authorities, and a range of welfare agencies such as the American Red Cross. Detailed, accurate reports of such responses are difficult to find. An excellent example (describing the events in Glendora during the 1969 storms) is given in a Los Angeles County Fire Department (1969) report. Although the procedures adopted by official agencies are precise and clearly integrated into the general disaster plan, there is

Table 3.2 Summary of emergency assistance during storms in Los Angeles County (from *Los Angeles County and cities disaster relief manual* 1970).

Agency	Rôle (in brief)
County	
1. Disaster Services Coordinator	Responsible for coordination
2. Sheriff	Law enforcement and public assistance
3. Fire Department	Communications and public assistance
4. Flood Control District	Monitoring, operational facilities management
5. Department of Hospitals	Ambulance and hospital facilities
6. Department of Public Social Services	Emergency welfare services
7. Health Department	Disease control, garbage disposal
8. Chief Medical Examiner–Coroner	Fatalities
City of Los Angeles	Disaster coordination and aid after emergency declaration and police communication
	Local within-city responsibilities (several agencies)
State	
1. California Disaster Office	Disaster coordination and aid after emergency declaration
2. Military Department	Emergency services
3. Highway patrol	Highway emergency management
4. Department of Human Resources Development	Provision of required skilled personnel
5. Department of Public Works	State highway protection and emergency engineering
6. Department of Water Resources	Remedial measures, repair and restoration of property

Federal	
1. USDA	Soil conservation, forestry protection
2. Department of Defense	Corps of Engineers fight floods, repair and survey damage
3. US Department of Health, Education and Welfare	Medical assistance
4. US Coast Guard US Federal Aviation Authority US Federal Communication Commission	Render equipment and facilities as appropriate
5. US Small Business Administration	Disaster loans at 3% interest to qualified groups
6. US Department of Housing and Urban Development	Insurance of damaged property
Non-governmental	
1. Red Cross	Emergency and recovery assistance to victims
2. Salvation Army	Provision of emergency food, clothing etc.
3. Association of General Contractors of America	Provision of construction equipment at cost
4. Engineering and Grading Contractors Association	Provision of personnel and equipment at cost
5. National Defense Transportation Association	Surveys of damage, transport facilities
6. Civil Air Patrol	Search and rescue
7. County Medical Association	Medical assistance
8. Hospital Council of Southern California	Medical co-operation

inevitably much improvisation that reflects the circumstances of the crisis, the flexibility required in dealing with the suffering public, the limited resources of public agencies, and the fact that the responsibilities of public agencies on private property are restricted (Lipkis *et al.* 1981). If public agencies cannot help private distress, the public may turn increasingly to organised groups such as the Tree People (Lipkis *et al.* 1981).

Damage to individual properties is one problem requiring emergency response that deserves special mention, because procedures are required for surveillance of many sites in the interests of public safety. These procedures vary between agencies, but those of the City of Los Angeles Building and Safety Department are amongst the most advanced. On receipt of a property damage message, the department surveys the property and prepares a short report. (These reports were analysed for 1969 in Part 2.) Then the authority may declare a house unfit for occupation (City of Los Angeles Building Code, section 91.0308) and/or issue an order to comply with building code provisions, listing the proposed necessary corrections (Code section 91.0103). The corrections involving slope failure often carry the additional requirements that the work should be carried out under the supervision of an approved soil-testing agency (City of Los Angeles, Building Code, section 91.3006f), or that corrective grading plans should be submitted together with a written report by a licensed soils engineer, and possibly also by an engineering geologist. When the work is completed and approved, the notices are rescinded.

MANAGEMENT OF SLOPES, WATER AND SEDIMENT

The management of water and sediment discharge is primarily in the hands of the LACFCD and the USACE, and their emergency operations developed locally over many years are codified and largely successful. During storms, the storm-related functions of the LACFCD are reconstituted as a Storm Organisation whose primary responsibility is to ensure that all facilities operate properly and are maintained to reduce flood damage to a minimum and to afford maximum water conservation (LACFCD 1964). (Water conservation is a very important activity of the LACFCD, and is achieved largely through re-charging above-ground and subterranean reservoirs. It has been ignored in this study only because it is of marginal interest to the geomorphological hazards under discussion.) There are two operations groups: one forecasting events and operating district facilities (such as dams, pumping plants, debris basins), the other effecting flood protection through maintenance and repair of channels and dams, and emergency assistance in flooded areas. The district recognises three magnitude levels of storm severity, each of which is associated with a prescribed set of responses (LACFCD 1976).

Activities of the USACE are based on delegated responsibilities under PL 84–99, and on other responsibilities that arise if a federal disaster is declared (for example under PL 93–288, The Disaster Relief Act). Specifically, the

USACE can assist during floods by furnishing technical advice and assistance; providing flood-fighting materials; moving equipment and operators for flood fighting; removing obstructions to flows and protecting emergency works (USACE 1978a); and by managing flood-control reservoirs and channels. The corps also plays an important rôle in costing damage and determining cost effectiveness of flood activities, and in reporting on storm and flood problems. The corps, like the LACFCD, recognises three levels of flood emergency in the Los Angeles region: *Category A* – all major floods in large drainage areas with actual or potential loss of life and serious damage; *Category B* – relatively localised floods (e.g. 'flash floods') which cause high damage and loss of life; *Category C* – floods in large catchments which, although not themselves serious, create conditions favourable for subsequent flooding.

Emergency flood operations are controlled through an Emergency Operations Center which includes groups concerned with flood-fighting, flood-control investigation, technical assistance, information and reports. Critical to the success of both the corps' and the flood-control district's emergency activities (and indeed all other emergency action) are efficient communications. A key part of the facilities, therefore, is the Los Angeles Basin radio system, and the telemetry system with its 43 telemetry stations throughout the southern part of the county (USACE 1978a).

The combined activities of both agencies during a serious flood can be exemplified by their work during 1969 (Simpson 1969; USACE 1969). The LACFCD controlled its own dams, and the USACE manned its flood-control basins in cooperation with the LACFCD, both parties paying particular attention to inflow–outflow rates and storage capacity, and debris-clogging problems. The LACFCD also operated 14 major pumping plants, maintained power and telephone and radio communication systems, and used its water-spreading grounds for water conservation. The USACE (acting under PL 84–99) additionally aided flood fighting, flood emergency preparation rescue operations, and especially the repair and restoration of damaged flood-control works (e.g. a stabiliser machine in the approach to the Santa Fé Reservoir on the San Gabriel River); and (under PL 81–875) assisted local authorities with rehabilitation work, including debris removal and bridge repair. In the aftermath of the two flood phases, much effort was expended on the key problems of debris removal from basins and of engineering works in anticipation of the next event.

Whereas the management of water and sediment discharge is integrated and controlled by only two principal agencies, the same is not true of slope erosion and slope failure away from watercourses, where management normally rests with the local authorities within which the failures occur. As a result, the responses are much more varied; they are usually incorporated in building codes (see p. 147). The procedures of the City of Los Angeles cited above (p. 130) exemplify the emergency response to slope failure on private property.

Paying the price: relief, insurance and litigation

BEARING OR SHARING THE LOSS

When property is damaged or business is disrupted by storms, who pays? The answer to this question in Los Angeles County is complex and is intimately bound up with changing legislation. Fundamentally, there are three choices for the individual beyond that of bearing the loss directly: to trust to the generosity of tax-raising authorities, to take out insurance in anticipation of damage, or to seek redress through litigation.

Trusting to the generosity of tax-raising authorities involves risk, because the nature and extent of loss-bearing by local, state or federal agencies depend *inter alia* on the uncertainties of political discussions, on the perception of the problem's magnitude, and on the extent to which claimants meet any conditions laid down. If the event is minor and can be managed by local agencies without recourse to state or federal agencies, the damage to an individual's property or business may have to be paid for by the individual himself, and his chief security may lie in having purchased insurance in advance (if it is available). A major difficulty in Los Angeles County (as elsewhere) has been that many people are unaware of the insurance possibilities, and even more are unaware of the need. Mrs De Pompa, a resident in Sunland, is typical. Her home was destroyed in the 1978 storms, but her insurance agent had told her she had all the insurance she needed, and she never thought she would have flood problems (*Los Angeles Times*, 19 February 1978). In such circumstances, some local subsidy through, for example, free police assistance, advice from building officials and debris clearance activities may be the only help available.

If the storms are sufficiently serious to attract the declaration of a local, state or federal 'emergency', then the situation changes dramatically – as it did in 1969 and 1978. Several laws are fundamental. First, PL 91–606, The Disaster Relief Act of 1970 (an act with several predecessors and successors, including especially PL 93–288, 1974), authorises federal assistance to state and local governments in federally declared disaster areas on a basis of applications ('damage survey reports') submitted through the state's Office of Emergency Preparedness to the Federal Office of Emergency Preparedness (known since 1979 as the Federal Emergency Management Agency, FEMA). This assistance, which includes repair, restoration, debris removal, temporary housing and environmental protection, extends to all public facilities, except streets and roads which are dealt with under Title 23, sections 120(f) and 125 of the US Code by the Federal Highway Administration. State assistance is available under the Emergency Flood Relief Law, which is similar in its provisions to the federal laws. For the private citizen, the federal legislation can bring assistance through tax relief and, in particular, through grants and loan guarantees administered, for example, by the Small Business Administration (PL 93–288, 1974, section 804). In addition, PL 93–288 makes provision for disaster preparedness and disaster recovery assistance, and allows for the full reimbursement to local agencies of the direct costs of disaster-related work.

From the point of view of geomorphological hazards, PL 93–288 included an important change in disaster-relief legislation. It incorporated, for the first time, landslide and mudslide into the definition of an emergency, in addition to storm, flood and other longer-recognised problems. Thus slope damage may now qualify for relief provided it is within an emergency area.

INSURANCE AND LITIGATION

Insurance One way of planning for hazard losses is to invest in insurance. The provision of insurance for flood and slope failure has a long and chequered history in the United States, a history littered with bankrupt companies and complex disputes (Langbein 1953, Kunreuther 1968). At present, flood insurance is governed federally by the National Flood Insurance Program established under the National Flood Insurance Act (PL 90–448, 1968), and other acts including the Flood Disaster Protection Act (PL 93–234, 1973; USDHUD 1974). The programme is administered by the USDHUD (Federal Insurance Administration), and subsidised insurance has been available to homeowners in the Los Angeles area on an initial 'emergency' basis since 1970. Unfortunately, although the number of policies taken out increases after each storm, the total number of subscribers by the time of the 1978 storms was disappointingly small (*Los Angeles Times*, 14 February 1978). In order for individuals to qualify for the 90% federal subsidy on premiums for buildings and contents, the community has to take positive steps towards minimising losses by implementing and enforcing land-use and flood-control measures and by discouraging development in high-risk areas. The insurance programme in Los Angeles County (as elsewhere) involves two phases: an initial emergency phase, for which *flood hazard boundary maps* are the basic documents; and the subsequent regular programme, for which *flood insurance rate maps* (FIRM) are prepared. The hazard zones on these maps form the basis for the determination of insurance premiums. Each map is accompanied by a report which describes the principal flood problems, the existing protection measures, the hydrological and engineering methods used to determine flood level–magnitude/frequency relations, and the application of the results of these methods to flood insurance, mainly through depth–damage curves. The research is usually subcontracted, for example to the USGS, the USACE, the LACFCD, the USDA or private consultancies (USDHUD 1977b). In Los Angeles County, the prediction of flood hazard for insurance purposes is not straightforward – because of the nature of ephemeral debris flows – and, as a result, special local attention is being given to this problem (LACFCD 1979b).

Once a community has agreed to participate in the Flood Insurance Program, individuals are required to insure within a specified period (originally one year) in order to qualify for disaster assistance. Because the preparation of flood-rate maps takes time, Los Angeles County had an interim programme for unmapped areas in which premiums were uniform; in mapped areas, the premiums vary with risk. The programme is expected to be completed during the 1980s. Once the community is on a regular programme, insurance is required by law in order to get financing to buy, build or improve properties located in USDHUD identified flood-prone areas.

Insurance against landslides and other slope failures is a quite different problem from that of flood insurance, and so is its local history. Insurance against landslides by private companies was temporarily terminated in the area in 1957–58, following the Portuguese Bend landslide. It was resuscitated in 1973 by the *Risk Analysis and Research Corporation* which sensibly employed the services of an experienced local engineering geologist, Charles Yelverton. The company's rôle was modified, at least in part, by the federal insurance provisions for mudslides in the National Flood Insurance Program (1974).

The national programme, which seeks to spread the burden of local slope-failure losses nationally and reduce costs by insisting on conditions for local-authority management, is not without its problems. Central to these is the urgent need for a precise method of predicting slope-failure hazard comparable to that for predicting flood hazard: the problem is very difficult, but some attempts are being made (see p. 137). A second problem is that the insurance only applies to failures related to accumulations of water on or under the ground. The failures must, it seems, be related to flooding before insurance relief can be claimed (Sorensen *et al.* 1975).

Litigation Litigation is an alternative means of alleviating the cost of hazard damage. And it is one that is becoming increasingly popular as the notion of *caveat emptor* fades and the concept of strict liability is accepted. To recoup property losses through the law requires the aggrieved party successfully to attribute responsibility for the damage. There is never a shortage of potential culprits – local authorities, developers, construction engineers, consultants and real-estate agents. And there is no shortage of legal arguments: those relating to negligence, negligent misrepresentation, concealment or fraud (Patton 1973) are possibilities. Geomorphological opinions are frequently called for in such arguments, and the flames of the law are easily fuelled by the central problem that predicting hazard damage is an imperfect art and equally competent geologists may have different but equally valid interpretations of a given situation. The concept of strict liability, which is based on the view that it is in the public interest to protect the consumer from dangerous mass-produced products, has come into use more recently, especially since a state supreme court ruling in 1963 (Greenman vs Yuba Power Products Inc., in Adelizzi 1969), and the extension of the concept to real estate (e.g. Connor vs Conejo Valley Development Co., in Adelizzi 1969). Litigation relates primarily to private property damage, and gives rise to what Patton (1973) called 'the

Southern Californian Landslide Litigation Syndrome'. Litigation consequent upon the Portuguese Bend landslide is a good example, for it ultimately cost over $1.0 million (Yelverton 1973, Albers vs County of Los Angeles, in Larson 1966). Professionals are common targets, and one of *their* responses is to seek insurance cover for themselves. Local authorities may also be liable even if, for example, a property is developed according to existing regulations but still fails (e.g. Sheffet vs County of Los Angeles, in State of California, Resources Agency 1971).

THE NEED FOR EDUCATION

Critical to the achievement of reduced storm, slope failure and flood damage and of reduced insurance premiums, is the need to educate the public about the nature of the geomorphological dangers they face. Those who own property are responsible for its maintenance, and it is clear that much hazard damage arises from poor property maintenance. On hillside sites, preventable damage may be caused, for example, by neglected drains and gutters, over-irrigation, and within-plot landscaping. Scullin (1966) found that as much as 95% of the storm damage in the City of Glendale in 1962 arose from lack of maintenance. Insurance agencies usually provide advice. For example, USDHUD publishes a brochure, *How to read an insurance rate map* (1977a), and the Risk Analysis and Research Corporation issues a leaflet on *Residential land maintenance* which the policy holder has to accept and display. Local and regional public authorities issue similar good advice. The City of Los Angeles Department of Building and Safety has prepared a sheet on the *Do's and don't's for hillside homeowners*, together with two more substantial guides, A *Guide for prospective buyers of hillside homes* (1978b) and the *Guide for erosion and debris control in hillside areas* (1978a), both of which explain for the layman the kind of precautions necessary to reduce or to provide protection from geomorphological hazards. Similarly, the LACFCD issues a *Homeowner's guide for debris and erosion control*.

But the mere preparation of such material does not itself reduce damage. It is difficult to get the information into the hands of property owners who need it, and (even if they receive it) there is little certainty that they will either understand or use it. Such propaganda is most effective at times of storm crisis, when minds are powerfully concentrated on a specific event. Also, response to education efforts appears to improve as the hazard experience of the community is extended. But the need for education is heightened in Los Angeles County because of the frequency with which residents change houses and the generally high rate of in-migration which brings new, often unsuspecting and environmentally innocent residents into the region. In these circumstances, the rôles of press and television are extremely important: the former, mainly through the *Los Angeles Times* and local newspapers such as the *Daily News*, certainly influences the better educated; the latter may help the less privileged sectors of the community as well.

Improving understanding: evaluation and prediction

FUNDAMENTAL PROBLEMS AND NEEDS

The fundamental difficulty of evaluating and predicting the nature of geomorphological hazards in Los Angeles County was explored in Part 1: the physical system itself is extremely complex and constantly changing, and many of the events occurring within it are relatively unpredictable in space and time. Attempts to evaluate and predict, essential prerequisites for efficient management, therefore require constant reassessment and improvement.

The growing recognition of these needs has led to an improvement over the years in the collection of environmental data. For example, Table 3.3 shows the increase in the number of sites at which rainfall, discharge of water, and sediment accumulation were measured in Los Angeles County between 1914 and 1969. Data from the present substantial network of stations give an admirable statistical basis for evaluating and predicting discharge of water and sediment in the channel system.

Slope erosion and slope failure have not been so well served, mainly because of the problems they pose for evaluation and prediction. They are essentially site problems in which each site reflects a unique interaction of numerous complex variables. The number of potential failure sites is infinite, and both natural and artificially engineered surfaces are affected; and sudden failure is normally involved at locations which are difficult to anticipate. Furthermore, a fundamental distinction between slope and channel problems is that the former are the concern of many different agencies throughout the administrative hierarchy, whereas the latter are the primary concern of two specialist agencies

Table 3.3 The growth of hydrological and sediment data monitoring stations in Los Angeles County.[a]

	1914	1938	1969
rain gauges	15 (40)[b]	143[c]	365[e]
stream gauging stations	4 (23)[b]	17 (43)[c]	49[e]
debris basins (reservoirs) – debris monitoring	0 (0)	17 (14)[d]	72 (14)[e]

Notes

[a]Because information for different dates is not strictly comparable, this table should only be regarded as giving a general impression of the change in monitoring stations over time.

[b]Grover *et al.* (1917) reported data from 40 gauges, of which approximately 15 were in Los Angeles County. Four reliable gauging stations were operating, and discharge was estimated using water levels at a further 23 sites in the San Gabriel and Los Angeles watersheds.

[c]Troxell *et al.* (1942) recorded 143 rain gauges (approx.) in Los Angeles County operated by the LACFCD, the USACE, the US Weather Bureau and other federal agencies; over 300 rain gauges were recording rainfall in the county. 'Basic discharge' records, reliably determined chiefly from reservoirs, came from 17 sites; discharge was calculated from 43 other sites.

[d]LACFCD (1938).

[e]Simpson (1969): figures refer to stations used in his report; the stream gauging stations constitute a representative sample only.

(the LACFCD and the USACE) that include their own monitoring, evaluation and planning divisions.

SLOPE PROBLEMS: MONITORING AND PREDICTION

Monitoring Monitoring of slopes is important in two main contexts: as the monitoring of slope erosion to assist the prediction of a sediment supply to channels under different conditions; and as the monitoring of landslide movement. Slope erosion is not well monitored. Only the USDAFS has provided long-term data (Krammes 1960, 1963). As a result, many estimates of soil erosion are based on extrapolations from their monitoring locations. While the sediment-yield record has come to provide the basis for many sediment-yield predictions (see p. 140), it is of relatively little value in the assessment of slope erosion at specific sites. There is still much to be learned about erosion rates on vegetated, burned and artificially cleared surfaces, and this is an important area for future research.

Monitoring of landslides is also rather poorly developed. The most significant contribution has been made by the City of Los Angeles Public Works Department (geology section), which has installed slope indicator wells, and water-level observation wells at 12 sites (J. Fitton, personal communication 1973), principally in the Pacific Palisades–Santa Monica Mountains foothills area. Other examples include the monitoring of the Portuguese Bend landslide by the Los Angeles County Engineer (Merriam 1960), and monitoring of the 1978 Malibu rockslide (Forsyth & McCauley 1982). There are several reasons for the paucity of landslide monitoring: in particular, it is expensive and requires personnel to make regular observations; and its main relevance is in the understanding of established, known slope-failure problems, being of less value in predicting the location of future failures.

Prediction Prediction of slope erosion and slope failure cannot yet wholly depend on empirical data derived from field monitoring at rather few sites. It depends in part on the cumulative experience of engineers and geologists derived from appraisal of sites before and during urban developments and after slope failure. Post-mortem studies of slope failure undertaken for litigation and insurance purposes (Leighton 1972) are particularly relevant to predicting future failures. Prediction is also based on the application to field conditions of established theoretical knowledge, and on the assessment of explanatory variables such as slope, rock structure and lithology.

The prediction of landslide hazard zones and relative slope stability by means of assessing environmental explanatory variables is relatively advanced in the Los Angeles region compared with elsewhere, mainly because the need for such assessment is so strong. In addition to the academic challenge posed by the problem of prediction, there is need in terms of actuarial assessment by insurance companies and the Federal Insurance Administration, of development potential by builders and engineers, and of hazard potential by planners and others concerned with land-use zoning.

The need is for two types of survey (Sorensen *et al.* 1975): for site-specific surveys, related to the evaluation or choice of specific sites for development, and for regional prediction.

Site-specific evaluations are normally carried out by an engineering geologist, perhaps working with a soils engineer. As Leighton (1966) pointed out, a trend in recent years has been for relatively more geological effort at sites to go into preventing rather than repairing failures. This is mainly because of growing demands for greater hillside safety, the realisation that prevention may be cheaper than cure, and the increasing recognition of the facts that landslides can be identified before development and that many failures are caused by the development process itself. In Los Angeles, the work is commonly done by consultants for private companies or public agencies, or by employees in local authorities. It normally involves detailed assessment of aerial photographs, available geological literature, and a wide range of basic data, mapping and subsurface studies in the field, and laboratory analysis of materials (Leighton 1976). It requires site evaluation of all the relevant factors, including especially historical evidence of slope failure, drainage conditions, subsurface characteristics and material properties, using a wide range of established and evolving techniques. It is the stuff of which are made Environmental Impact Statements (prepared under the National Environmental Policy Act, 1969, and subsequent legislation) and investigative reports to city building and safety departments (City of Los Angeles 1972a, Building Code, section 91.2804). Detailed geological site studies may be undertaken – indeed they are increasingly required – at several stages during the development process. Information from such studies may be included in reports prior to purchase, in planning and design reports, in tentative tract reports, in bulk grading and grading plans, in in-grading inspections, in final grading reports, and in storm-damage reports (Leighton 1966).

Regional prediction, designed to identify areas of slope-failure hazard, is a less mature art, but it has nevertheless been extensively practised in the Los Angeles area by federal, state and local agencies. Most of these efforts are interpretations based on historical landslide data (in which the present distribution of past slope failures is used to predict future failures), on the mapping and integration of explanatory variables in the slope-failure system, or on a combination of both these approaches.

There are several examples of regional landslide maps for Los Angeles County prepared by federal agencies. These include a state-wide reconnaissance map prepared for a US Coast and Geodetic Survey study of the potential consequences of a major earthquake in California (Radbruch & Crowther 1973). It does not attempt to show relative safeness for construction; rather, it shows estimated relative amounts of area covered by landslides, and is based on a combination of systematic analyses of some variables (e.g. slope, precipitation, geological units) and subjective interpretation of others. Also a federally sponsored project, Campbell's (1973) isopleth map of landslide deposits in the Malibu coast area is an original attempt to summarise landslide distribution in a way that may be useful in comparing the relative slope stability of different areas and in serving as a guide to the location of potential slope-failure problems using only an inventory of recorded slope failures. The map can be combined with other quantified map data for the preparation of higher-order derivative maps, and it may be more suitable for some regional planning purposes than simple inventory maps (Wright *et al.* 1974).

The state has also been active in regional slope-failure prediction surveys (Blanc & Cleveland 1968, Cleveland 1971a, Evans & Gray 1972). This work has both state-wide and local dimensions, and is related, in part, to the requirements of USDHUD for actuarial data. Cleveland's (1971a) method is based on theoretical understanding of landslide processes, the study of landslide distribution, and the analysis of the distribution of background factors (such as rock-vegetation factors and geomorphology), energy factors (including climate, weather, and especially precipitation) and special factors (such as adverse geological structures and high water tables). Cleveland's (1971a) broad-brush regional prediction was developed locally in Ventura County by Evans and Gray (1972), where three major 'mudslide risk zones' were mapped, and a series of key factors was used in classifying terrain. (Mudslides are defined by the USDHUD as 'movements down a slope of a mass of natural rock, soil, artificial fill or a combination of these materials caused or precipitated by the accumulation of water on or under the ground' – Evans & Gray 1972.)

City and county agencies have also been concerned with mapping slope-failure hazards. For example, the City of Los Angeles Department of Public Works (C. A. Richards, personal communication 1973, 1979) is preparing relative stability maps for especially hazardous districts which distinguish among relatively unstable, relatively stable, and stable areas on the basis of the experience of the department's engineering geologists and all other sources of geological information. Such maps are general guides of value in planning and in determining where additional geomorphological information is desirable. More simply, the city's 'hillside areas' – areas in which slope-problem control ordinances are effective (City of Los Angeles, Building Code, section 91.0403) – successfully encompass land in which slope problems occur, although the regional boundaries follow cultural features such as roads and city limits (Palmer 1976).

Los Angeles County is covered (wholly or in part) by several other landslide surveys. Leighton (1966) prepared a preliminary map of landslides (on natural and artificial slopes) using all available historical information; McGill (1959, 1973) field mapped landslides in the Pacific Palisades; and Morton and Streitz (1969) identified landslides in the San Gabriel Mountains from aerial photographs. These surveys are based on different data bases and, while not in themselves predictors of the location of future failures, clearly inform judgements of slope stability.

A final category of slope-hazard prediction deserves mention. Those concerned with geomorphological problems in the region recognise that brush fires may exacerbate slope-erosion and sediment-yield problems. Thus, erosion potential reports (which evaluate the problems and propose solutions) are commonly prepared following a fire, by the LACFCD or some other county or city agency. For example, the study of the erosion potential in the City of Rolling Hills following the Crenshaw fire in 1973 (Los Angeles County Department of County Engineer 1973) included a map which assessed the sediment yield for the 25-year storm for each affected catchment on the basis of field study of soil thickness, slope, vegetation, and other factors. Such surveys commonly employ the findings of regional sediment-yield studies, such as that of the LACFCD (discussed in the next section), and recommend protective action to limit debris yield (LACFCD 1970a).

The variety of efforts to predict slope stability and identify landslide hazard areas reflects the need, the difficulties of the task, and the range of possible approaches. Some of the efforts in Los Angeles are highly innovative and have greatly improved understanding of where failures might occur. The regional surveys and the experience of those involved in them contribute not only to local planning requirements, but also to state-wide and national studies (Baker & Chieruzzi 1959, Krohn & Slosson 1976, Leighton 1976, Fleming & Taylor 1980).

FLOODING AND SEDIMENT PROBLEMS

From the geomorphological point of view, the prediction of flooding and sediment problems has both an upstream and a downstream component. Upstream, the primary aim is to predict the quantity of sediment and water that drainage catchments will produce; the main downstream objective is to predict where the water and sediment might go, and what damage they might cause.

Upstream: predicting sediment yield Predicting sediment yield from upstream catchments has been of major importance for many years because it is an essential prerequisite for the efficient planning of debris-collection structures, such as debris basins. The predictions help to determine where such structures should be placed, how big they should be, and how long they will last. The main efforts at prediction use sedimentation data derived from debris basins to establish relationships with easily quantifiable environmental variables of predictive value.

Methods of sediment-yield prediction are not generally explicit in the early plans to design and locate debris control structures. At best, it seems that estimates were made from sediments deposited by floods in quarries or on alluvial fans. Certainly there seems to have been no systematic monitoring of sediment traps (Eaton & Gillelen 1931). The 1934 floods themselves provided valuable evidence that was used in designing the first major debris basins in the La Cañada Valley.

The transition between early estimates and modern, empirically based predictions of sediment yield is made in the important study by Retzer *et al.* (1951) on the origin and movement of sediments in the Los Angeles River watershed. The importance of this study lies in the fact that, although it does not measure soil loss or predict sediment yield, it precisely describes and analyses for the first time many of the physical properties of the region (using a sample of 46 small catchments), and it establishes several important physical relationships. The study paves the way to more precise, quantitative measurement and analysis of geomorphological variables (Maxwell 1960), and to precise predictions of sediment yield; but it stops short of that goal. A further substantial and fundamental step towards the precise prediction of sediment yields in general, and the effect of fire on erosion rates in particular, was made by Rowe *et al.* (1954) for the USDAFS. Their detailed analysis of available precipitation, runoff and reservoir siltation data in the mountains of southern California incorporated recent progress in both scientific hydrology and geomorphology and led to some important quantified generalisations (e.g. Fig. 1.14).

The USDAFS study was followed by the LACFCD's (1959) similar, fundamental assessment of sediment yield, in which a predictive equation for sediment yield was developed for the check-dam and channel-stabilisation programme. A large suite of variables was evaluated, and finally reduced to the following prediction equation:

debris production rate (F) (in cu. yd/sq. mile) =

$$\frac{35\,600\, Q^{1.67}\, Rr^{0.72}}{(5 + V.I.)^{2.67}}$$

where

- Q = peak runoff in cu. feet/s per sq. mile resulting from the maximum 24-hour rainfall of a given storm;
- Rr = relief ratio (ratio of total vertical dimension to the horizontal distance in the watershed parallel to the main channel) (this variable effectively represents several others such as channel and basin slope);
- $V.I.$ = vegetation index, an index based on type and cover, which accommodates the effects on sediment yield of fire history.

Values were modified after the calculation was complete in terms of antecedent rainfall conditions. In order to test the accuracy of the equation, debris production for each storm in each of 12 basins between 1935 and 1955 was estimated using the equation, and then compared with debris production records: the estimated volumes differed from the actual volumes by only 2%. The study, recognising the fact that sediment yield in the mountain frontal zones varied, described seven major regions with their distinctive characteristics; it also estimated sediment-production rates based on the 1938 storm (the largest on record) and the 50-year storm (used for design purposes). The classification and the prediction equation enable sediment yield to be predicted from any catchment with some confidence (see Fig. 1.23). This fundamental study is known as 'the blue book'.

Tatum (1963), of the USACE, developed a similar but independent technique to that of the LACFCD for predicting debris production. He recognized five factors that were fundamental in controlling yield: debris catchment area, slope, drainage density, hypsometric-analysis index, maximum three-hour rainfall occurring in a storm after prior rain has conditioned the ground, and the effect of burning. The technique involves estimating the ultimate debris potential index for one square mile in the maximum debris-producing area (i.e. the yield in a storm after a 100% burn), and modifying this value in terms of percentage corrections for the major factors (derived from graphs), and correcting for the actual area burned and the number of years since the burn took place.

The translation of sediment-yield prediction from catchments of known yield to those of unknown yield through the use of geomorphological predictors was significantly advanced in the relatively little-quoted study by Lustig (1965). Lustig created six geomorphological predictors, and established relationships between the values of these variables and sediment yield for several drainage basins. From such relationships it is possible to predict sediment yield for any other basin in the same environment by measuring the values of the geomorphological predictors and using regression analaysis.

The most recent and sophisticated research in the sequence of debris-prediction studies owes much to its antecedents. The study of sediment yield in the Transverse Ranges by Scott and Williams (1978) used debris-basin data, especially from the 1969 storms in Los Angeles County, and a wide range of predictors to estimate yield in ungrazed areas; it, too, was based on regression analysis. Of the 16 morphometric variables examined, elongation ratio (*ER*), drainage area (*A*), and area of slope failures (*SF*) were of the greatest value in regression analysis. Other variables included a fire factor (*FF*, 0–100, where 100 = 100% burn followed immediately by a storm), mean annual precipitation (*MAP*), a storm-precipitation factor (K = 10-day × (24-hr precipitation)^2), and total time of concentration (*TC*, time required for water to flow from remotest part of basin to outlet). The following regression equation exemplifies the relationships:

$$\log Sy = -0.981 + 1.132 \log A - 1.059 \log TC + 1.322 \log ER + 0.363 \log SF + 0.250 \log FF + 4.847 \log MAP$$

where *Sy* is sediment yield of the January 1969 storm in cubic yards for locations in the eastern Transverse Ranges of Los Angeles County. Scott and Williams' approach has been adopted in a preliminary assessment of sediment yield and its acceleration by off-road vehicle use in the Cañada de los Alamos Basin by Knott (1980).

The methods of predicting sediment yield in mountain catchments provide an object lesson in the progressive development of a geomorphological concept, and the results obtained from them have undoubtedly been of great value in planning the size and location of debris basins and in predicting erosion potential after fires. Morphological variables figure prominently in all the methods, mainly because of their predictive power, precision, and easy recovery from topographic maps. Also critical to all of them, but more controversial, is the use of historic sediment-yield data from older debris basins and reservoirs. The basis on which such data are calculated is often not clear. It seems that mostly (and especially for reservoirs) they are derived from profile and cross-section surveys done as soon as possible after the storms have passed (LACFCD 1938). Occasionally for some debris basins, sediment volumes may be estimated from 'truck loads' of sediment removed – a less accurate assessment method. Also, some of the data may incorporate both the reducing effects of debris-retention structures, like check dams, and the enhancing effects of erosion-vulnerable road cuts, firebreaks and other clearings; such effects change over time and are not explicitly incorporated into prediction models. There are still other gaps in the data relevant to the models: for example, little is known of the ratio between rate of weathering and rate of erosion.

The methods reflect such problems; they also reflect a response conditioned by time and money for research. The result is a compromise, robust, adequate and improving.

Downstream The prediction of water and sediment discharge on the plains beyond the mountains, and of the paths it is likely to follow, is largely based on the monitoring of flows at gauging stations, on flow estimation at surveyed channel cross-sections, and on the experience of actual flood events. Such data are used in well-established hydrological methods of estimating the physical characteristics of flows for given recurrence intervals. The present studies of flood hazard for federal flood insurance and the preparation of flood insurance rating maps (USDHUD 1977b) are the latest in a sequence of such studies that extends back at least as far as the reports on the 1914 storms (Los Angeles County Board of Supervisors 1915).

One study that deserves special mention is the LACFCD's (1979b) attempt to develop for insurance studies a methodology for mapping mudflow hazard areas and predicting peak flow rates and debris volumes. This problem is not quite as straightforward as it might seem: to predict the 100-year ('base') recurrence interval debris volumes is extremely difficult, and requires consideration and estimation of, *inter alia*, mean annual production rates, modifications to take the effect of fire into account, and prediction of peak flow rates. Once bulked flow rates are determined, debris-laden flows are analysed in the same ways as clear flows – at least as far as depth and velocities go. But more is required, because the location of deposition forms a major part of the hazard. Thus areas of potential deposition on alluvial fans have to be mapped, taking into account the peculiar geomorphology of the features and the occurrence of obstacles.

Anticipating problems: abatement and control

INTRODUCTION

Improved understanding and the ability to predict geomorphological problems allows management agencies better to anticipate the problems and to control or alleviate them through prior action. The most important ways of implementing such actions are through land-use zoning and subdivision ordinances, in which certain developments are proscribed in hazardous situations, and through building codes, in which site development and building construction are regulated at specified hazardous locations. The type of restrictions range from avoidance (e.g. preventing building on unstable sites) to engineering control (e.g. building dams and reservoirs). Engineering control of slope failure is largely embodied in building codes, although post-failure engineering remedies normally are not; engineering controls of floods and sediment are normally implemented through management-agency plans.

The comments in the following pages are based on the most recent documents available to the author. In some instances, the controls described may have been modified, deleted or supplemented, especially as a result of reviews since the storms of 1978. The comments here are also selective; it is not possible to discuss in detail all hazard provisions in all development-control documents.

CONTROLLING SLOPE FAILURE

Introduction As a result of state legislation, communities in California are now required to recognise landslide hazards in their planning and zoning ordinances and grading codes (Leighton 1976). Any community seeking to control slope failures within its administrative area is likely to ask a series of questions (Sorensen *et al*. 1975). What is the nature of the hazard? Where can it occur? To what extent could it occur? When is it a disaster? What can be done about it? What approach, method or procedure should be used? Who will be involved, what will they do and how will they work? How will the total effort be best coordinated? Who will pay the cost?

Critical to answering this multiplicity of questions is the community's scientific understanding and political perceptions of the problems, and its recognition of the range of acceptable solutions. The range of solutions open to a community has a firm geomorphological basis: fundamentally, attempts can be made to avoid urban development in hazard areas altogether, to reduce shearing stresses applied to slopes, to increase the shearing resistance of slopes, or to use some combination of these. Within each of these categories, many specific solutions are recognised (Root 1958, Kockelman 1980), of which those most used in Los Angeles County are shown in Table 3.4.

But, as emphasised previously, the integration of solutions into building and zoning codes is not simply a reflection of scientific assessment and engineering capability. It is also a tangible compromise in the legislative arena between developers and the building industry, seeking to maximise their opportunities and profits, and the community exercising what it perceives to be its social responsibilities. The form of the compromise inevitably varies from place to place and over time, and so it is within the 70 or so cities of Los Angeles County. The following sections assess some of the ways in which the range of control measures is incorporated into zoning and subdivision ordinances, and into building codes.

Zoning and subdivision ordinances The county and the cities within it exercise control over land use through zoning ordinances, which allocate particular uses to pre-determined zones. The design details of developments within these zones – such features as lot shape – are controlled through subdivision ordinances, which must conform to the state's Subdivision Map Act. Buried within these complex and fundamental planning documents, reference is usually made to problems of slope failure and flooding, and the zoning and subdivision maps are the primary documents in which implicit assumptions about, and perceptions of, such hazards are enshrined. The essence of the ordinance intentions, as far as geomorphological hazards are concerned, is to prevent certain developments in hazardous locations and to ensure that appropriate precautions are taken if permitted developments do take place in hazard zones.

The county zoning code (e.g. Los Angeles County Regional Planning Commission 1972), while it clearly recognises and responds to flood hazards (e.g. article 4, chapter 3; article 5, chapter 7) by prohibiting most public building in flood zones, says little directly about what it calls 'sloping terrain'; on the other hand, the county subdivision ordinance (No. 4478, section 94) requires geological reports to accompany tentative development maps if an existing or potential geological hazard is deemed to exist, places constraints on lot size in sloping terrain, and reserves the right to disapprove or require protective improvements to developments on land subject to flood or geological hazard (section 158). Indeed, since 1962 'no permits, subdivisions or projects are approved on known geologic hazardous areas' (Los Angeles County Engineer 1965, Los Angeles County Building Laws 1968, section 308).

Table 3.4 Principal methods of preventing and controlling slope failures in Los Angeles County.

Major solution (After Root 1958)	Specific methods	Principal adopting codes
Avoidance	I. *Control of location, timing and supervision of development*	
	A. Land-use restrictions, recognition of hazardous regions, subdivision controls	CLABC, CBL, UBC
	B. Engineering geology and/or soils surveys before, during and after development	UBC, CLABC, CBL
	C. Seasonal limitations to slope development	CLABC, CBL
	D. Relocation, or avoidance of failed sites	CLABC, CBL
Reduce shear stress	II. *Control of cut and fill*	
	A. Limit/reduce slope inclinations, cut and fill	UBC, CLABC, CBL
	B. Limit/reduce slope-unit lengths	UBC, CLABC, CBL
	C. Remove unstable material	a
	D. Expansive soil provisions	CBL
Reduce shear stress and increase shear resistance	III. *Improve drainage*	
	A. Surface	
	1. Terrace drains	UBC, CLABC, CBL
	2. Drains	UBC, CBL, CLABC
	3. Others	UBC, CLABC
	B. Subsurface	
	1. Drains	UBC
	2. Drain wells	
	C. Control irrigation	CBL
Increase shear resistance	IV. *Retaining structures*	
	A. Buttress at slope foot	CLABC[a]
	B. Cribs or retaining walls	UBC[a]
	C. Piling, tie-rodding or other foundation engineering controls	CLABC[a]
Chiefly increase shear resistance	V. *Protect surface*	
	A. Control vegetation cover	UBC, CLABC, CBL
	B. Harden surface (e.g. concrete cover)	CLABC[a]
	VI. *Compaction*	
	A. Control fill compaction	UBC, CLABC, CBL

[a]Methods not often incorporated in building codes.
UBC = Uniform Building Code.
CLABC = City of Los Angeles Building Code.
CBL = Los Angeles County Building Laws.

The City of Los Angeles Planning and Zoning Code (1972b), which includes the zoning and subdivision ordinances, makes similar provisions: its designated hillside area (section 91.0403; see also Fig. 2.9), in particular, is a clearly described hazardous zone in which certain regulations (mainly in building codes, see below) are enforced. In many other cities, too, hazardous slope areas are defined, and the ordinances register the authorities' right to take special action or to implement special building code regulations.

Subdivision development permission rests essentially with powerful subdivision committees which must approve tentative tract maps. The committees may include an engineering geologist (as in the City of Los Angeles), normally have powers to require appropriate geological reports, and usually expect at least professional geological appraisal of proposals (Los Angeles County Engineer 1970, Yelverton 1972). The opinion of geologists may be required at the tentative tract map stage, at the grading plan stage, after grading is completed, and after construction. The contents and standards required of geological reports are now commonly specified (Los Angeles County, no date; City of Los Angeles, no date); other aspects of building code control are examined below.

Control through building codes Ground engineering and building within a subdivision are controlled mainly through building codes. These codes are designed to prevent damage, and they attempt to achieve this aim through the establishment of design and construction standards and the provision of procedures of development control. The codes vary considerably, both spatially, within Los Angeles County, and temporally, within individual administrative areas.

The *spatial variability* of building code provisions is revealed by an analysis of the codes and the results of a questionnaire survey sent to all cities in the county in 1973. Six major classes of code (Fig. 3.3) are identified: (a) Uniform Building Code (UBC); (b) UBC with local modifications; (c) own ordinances; (d) county ordinances (unincorporated land – CBL); (e) cities contracted with county for some or all slope-related services; (f) as (e) but with some powers reserved.

Of the 77 cities involved, 62 responded to the questionnaire. Of the 30 cities contracted with the country, 24 replied, and most of these had some or all their hazard-related services from the county; only 4 reserved some powers of their own. Of the 47 cities with their own ordinances, 38 replied. Of these, 21 depended on the UBC, 8 used the UBC with local modifications, and 9 had their own ordinances. In addition, slope-related developments by state or federal agencies, while controlled by local regulations, may be further restricted by special requirements, such as those imposed by the Federal Housing Administration (National Academy of Sciences 1969).

The Uniform Building Code (the UBC; International Conference of

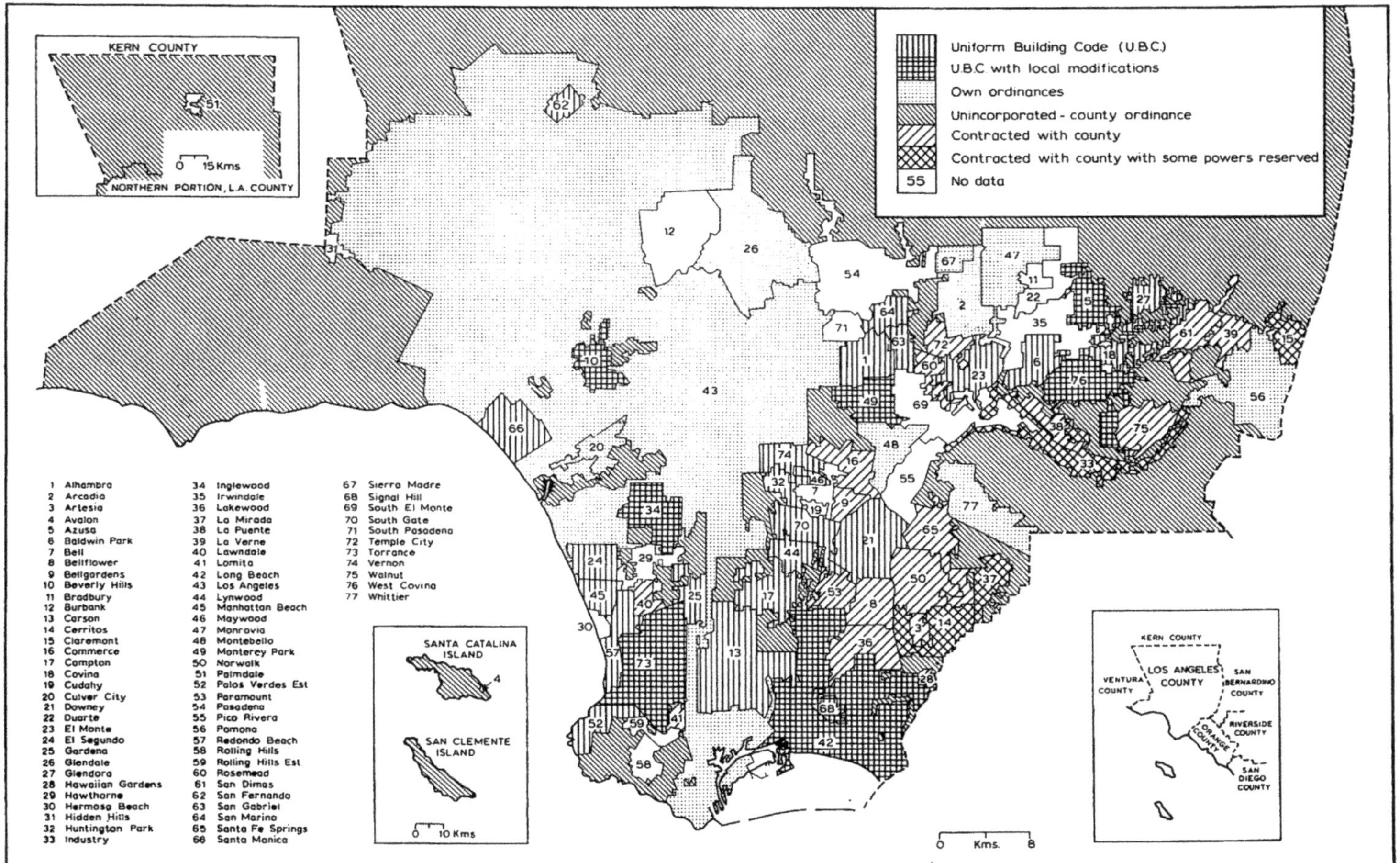

Figure 3.3 Spatial pattern of building codes in use in Los Angeles County, 1973. For explanation, see text.

Building Officials 1973) was first enacted at a meeting in Arizona in 1927. Chapter 70 of this code is concerned to regulate grading (i.e. slope engineering) on private property. It was only introduced in 1963, and was intended to be of universal application, despite the fact that it was based heavily on southern Californian experience in the previous years, and the southern Californian section of the Association of Engineering Geologists played a leading rôle in its formulation (Scullin 1966). In addition to rationalising existing codes, one of its primary aims was to ensure adequate and coordinated professional supervision of grading work, mainly by soils engineers and engineering geologists. By 1973, Chapter 70 had been adopted (at times with local modifications) by at least 29 cities in the county. California state law now requires that all cities and counties use Chapter 70 as the minimum grading code for hillside development if they do not have their own equal or more stringent codes (Weber 1982).

The unincorporated land of the county – in fact, the largest administrative area – is covered by Chapter 70 of the County of Los Angeles Building Laws (CBL; Los Angeles County 1968), which has many elements in common with the UBC. (Unless otherwise stated, comments here relate to the 1968 volume.) The county code also applies to those cities that contract with the county for building services, dates of contract usually coinciding with the time of incorporation and ranging from the 1940s to the present. The City of Los Angeles is by far the largest autonomous city in the county. It is covered by the City of Los Angeles Building Code (CLABC; City of Los Angeles 1972a). These three sets of codes (the UBC, CBL and CLABC) in combination control slope development in over 80% of the county.

A detailed analysis of these three control documents and the building regulations in other cities reveals some of the diversity of local practices. In the following selective discussion, the categories of methods refer to those listed in Table 3.4.

Control of location, timing and supervision of slope development (category I, Table 3.4) is normally exercised through the zoning, subdivision and building codes and, especially, through the issue of grading permits. The issue of the permits may be withheld for specified hazardous locations or, more likely, will be qualified if slope problems are deemed to be significant. Some authorities outline hazardous areas in which special conditions apply. The hillside area of the City of Los Angeles (ordinances 129279, 129885, and 131994; see Fig. 2.9) is one example; the cities of Glendale and Montebello make similar designations; and some other cities, such as Torrance, define hillside lots in terms of slope thresholds. Other requirements vary considerably, but now commonly include provisions for engineering geology and soil reports *before* grading (e.g. CBL, UBC, CLABC), and some stipulate supervision *during* grading by approved geologists and others at specified times (e.g. CLABC, section 91.3002; UBC and CBL, section 7014). As a result of such measures, geomorphological advice (together with that relating to other aspects of geology and engineering) may be provided at up to seven stages in the development process (see p. 138; Leighton 1972).

The period of development (category IC, Table 3.4) is limited in some areas to exclude or severely restrict slope work in the rainy season. In CLABC, the rainy season extends from 1 November to 15 April (ordinance 151828); in CBL it is from 1 October to 15 April; in Montebello, from 15 November to 15 April; and in Sierra Madre, from 1 November to 1 May. This provision is not in the UBC. In CLABC, if site developments are not completed before the onset of the rainy season, control devices and protective measures such as de-silting basins may be required by 15 September (ordinance 135199; in the CBL, the date is 1 November, section 7017).

Both CLABC and CBL have sections in their grading ordinances specifically concerning slope-failure problems. In CLABC, section 91.3011 defines 'landslide', 'active landslide', 'historic landslide' 'prehistoric landslide' and 'possible prehistoric landslide', and gives details of restrictions on permission to construct buildings or to do grading work in areas of each type and associated 'questionable areas'. CBL section 308(b) discusses geomorphological hazards (including landslide, settlement or slippage), and the conditions under which permits may be issued for developments in areas affected by them.

The control of cut-slope and fill-slope inclination (category II, Table 3.4) is universally recognised to be important in reducing shear stress, but there are substantial differences in the maximum slope inclinations permitted. In the City of Los Angeles, where some of the severest controls have been developed, no cut or fill slope can normally exceed a 2 : 1 ratio – 50% – (section 91.3005/6). The UBC has similar provisions, and also restricts the natural inclination of slopes on which fill can be placed. In the county, the normal ratio is 1.5 : 1 (66.7%) for cut slopes and 2 : 1 for fill (ordinance 10088, 1970), and in some cities (such as Signal Hill and Culver City) the highly permissive 1 : 1 ratio (100%) limit for cut slopes was still used in 1973, but authorities had powers to restrict development if necessary. In the opinion of some experienced engineers, the 2 : 1 ratio is too generous for certain conditions, but reasonable for many hillside areas. The problem is, of course, to formulate a general rule that effectively covers most circumstances but which is not unnecessarily restrictive. In CLABC, there is a specific additional requirement that no slope will exceed bedding plane dip where the cut slope is on the dip side of the strike line (section 91.3005), a measure designed to prevent bedding-plane slip.

Another requirement is that slope lengths should not be too great, in order to reduce runoff hazards and increase strength. The UBC requires drainage terraces at least 6 feet (1.83 m) wide, at not more than 30-foot (9.14 m) vertical intervals (section 7012). In CLABC, no cut or fill slope can exceed 100 feet (30.48 m) vertical without horizontal benches at least 30 feet (9.14 m) wide at each 100-foot vertical interval; intermediate benches of at least 8-foot (2.44 m) width are required for both cut and fill slopes at 25-foot (7.62 m) vertical intervals (section 91.3005/6). In the CBL, on cut slopes exceeding 40 feet (12.19 m) vertical, at least 8-foot (2.44 m) wide terraces are required at 25-foot (7.62 m) vertical intervals, and on cut slopes over 100 feet (30.48 m) in height, a drainage terrace near mid-height not less than 20 feet (6.10 m) wide is required (section 7009); on fill slopes over 30 feet (9.14 m) vertical, terraces at least 8 feet (2.44 m) wide at intervals not exceeding 25 feet (7.62 m) are required, except that where the total slope height exceeds 100 feet (30.48 m), one terrace near mid-height not less than 20 feet (6.10 m) wide must be provided. The UBC, CBL, CLABC and some other city codes (such as Beverly Hills) include set-back requirements that control minimum distances between slopes, property boundaries and buildings.

Shear stress is reduced and shear strength increased (and landslide and soil erosion potentials are both thus reduced) if water is efficiently and safely removed from vulnerable slopes (category III, Table 3.4). Terrace drains, to catch slope debris and carry away water, are now almost universally required. The width and spacing requirements of the UBC, CLABC and CBL vary somewhat (as described in the previous paragraph). The codes usually also specify minimum gradient and depths for the drains, and some limit the tributary catchment area of a drain (e.g. UBC). Other drainage provisions vary. The UBC, for example, mentions subsurface drainage and provision for water disposal, whereas the CLABC provides considerable detail (section 91.3008) on such features as inlet and outlet structures, downdrains, subdrains and drainage maintenance.

Retaining structures (category IV, Table 3.4), such as buttress fills, piles and retaining walls, are not normally mandatory, but where they are required to increase shear resistance of slopes, their design is sometimes specified (e.g. CLABC, section 91.3010; section 91.28). According to Leighton (1966), buttress fills are the commonest and amongst the cheapest and most successful remedies for unstable cut slopes.

The shear resistance of slopes can also be increased (category V. Table 3.4) – and the potential for slope erosion and failure thereby reduced – if access of water to the surface is restricted by a vegetation cover, or by sealing the surface (as with concrete). Vegetation may, of course, also serve to bind the soil. The UBC is permissive in this respect (section 7013). The CBL (section 7013) requires a grass cover on cut slopes over 5 feet (1.52 m) high, and on fill slopes over 3 feet (0.91 m) high, and permissive combinations of trees, shrubs and grasses to standard specifications on slopes over 15-foot (4.57 m) vertical height, together with adequate and approved irrigation systems. The CLABC has similar (but more detailed) requirements to the county (section 91.3007); the Building and Safety Department provides a list of approved plants, and restoration of vegetation on graded slopes is required within 30 days after grading where no building permit is in effect (section 91.3001); furthermore, existing vegetative ground cover on slopes cannot be removed except under specific circumstances (91.3001).

A final important control of shear resistance concerns the compaction of fill (category VI, Table 3.4). The UBC, CBL and CLABC regulations require compaction to at least 90% of maximum density, and procedures adopted (although not specified in detail in the codes) include the ASTM Soil Compaction Test D1557–58T. Several cities (Los Angeles, Signal Hill, Beverly Hills and others), aware of recent storm damage to slopes engineered before 1952, require re-testing of compacted fill slopes built before that date if redevelopment is contemplated.

Many building regulations also control the response of property owners to slope damage. The procedure in the City of Los Angeles has been described above (see p. 130; CLABC section 91.0103); similarly, the City of Beverly Hills requires repairs within 90 days of receipt of an order, and specifies the consequences of failure to do so. The purpose of such requirements is to remove danger to public property and/or to isolate hazardous locations. The nature of corrective action may be specified in codes, but in general such action is likely to be determined by the specific circumstances and setting of the site, the professional opinion of geologists and engineers, and the relative costs of alternative actions. Remedial action is usually based on the same principles as those underlying the preventative controls in the codes (Smith 1966).

CHANGES OVER TIME. The complex spatial variations of slope-failure controls are matched by changes over time within different administrative areas. The City of Los Angeles provides an excellent illustration of the evolution of building code controls over private hillside development. Each landmark in this evolution (Table 3.5) normally involves a physical stimulus (such as storm damage or serious landsliding), and at least three phases of response: the establishment of a committee of enquiry as an immediate response; a professional response from the committee in the form of a report

Table 3.5 Examples of post-crisis responses in the City of Los Angeles to slope problems arising from storm damage.

1.	1952	*Superintendent of Building and Safety*'s investigation of slope failure, leading to 1952 grading ordinances
2.	1956	*ad hoc Geological Hazards Committee* established
3.	1958	*Engineering Geologists Qualification Board* established in the Department of Building and Safety
4.	1962	*Review committee* leading to grading ordinance revision, 1963
5.	1965	Mayor Yorty establishes the *Committee of the American Institute of Professional Geologists* on the *Geologic Environment in the City of Los Angeles* (report, American Institute of Professional Geologists 1966)
6.	1965	The mayor's *ad hoc landslide committee* (report, City of Los Angeles 1967)
7.	1978	*Storm Damage Task Force Report* (22 June 1978); and *Building Codes Committee Review* of *1978 Storm Damage* by the *Association of Engineering Geologists* (report, Association of Engineering Geologists, 7 August 1978)

For discussion of 1, 2, 3 and 4 see Jahns (1969).

and recommendations; and a political response to the recommendations that is ultimately reflected in local ordinances and code changes. Over time, the clearest tendencies have been to make acceptable types of control more severe, to make new concepts and controls acceptable, and to improve implementation procedures. In the first category, the initial grading ordinance (1952) focused on the control of grading through permits and inspection, and on specific design limitations of slope steepness, drainage provisions, and fill compaction. These controls have been strengthened in subsequent years. For example, cut slopes were limited to 1 : 1 gradients in 1952; after the 1957–58 storms, the maximum angle was reduced to 1.5 : 1 in the Santa Monica Formation and in certain other areas; and in 1963, the maximum angle of most cut and fill slopes was reduced to 2 : 1 (City of Los Angeles 1967).

There are other ideas that have become acceptable since 1952, and many have been introduced into codes. Examples include the definition of a hillside hazard zone, controls on hillside vegetation planting, and seasonal restrictions on grading activity. Once introduced, most of the ideas have been developed and improved.

Efforts to improve implementation procedures focus mainly on the methods by which tentative tract maps, grading plans and grading operations are approved and supervised. The changing rôle of geology and geologists in the procedures illustrates these efforts well. Although the 1952 ordinance recognised the importance of geological advice in grading control (Yelverton 1971), geological reports prior to the issuing of grading permits have only been required since 1956; controls on the qualifications of geologists were accepted only in 1958; guidelines for the contents of geological reports were first laid down in 1960; and additional responsibilities for geologists and others, including supervision and certification of grading and subsurface investigations, were introduced in 1963 (Jahns 1969, Richards 1970, Yelverton 1971). These and subsequent innovations not only reflect a need, they also reflect favourably on the substantial and growing influence of a small number of distinguished geologists in private practice and public service.

The City of Los Angeles led the way in establishing controls on slope development and in recognising the importance of engineering geology and soils engineering. The cities' 1952 codes were adopted or adapted, for example, in Beverly Hills (1952), Pasadena (1953), Glendale (1954), and Burbank (1954); and the county adopted grading codes in 1957, and a qualifications board in 1960 (Scullin 1966).

To adopt formal controls on hillside development in the form of building codes is only part of the solution to the problem of building safety on slopes. Equally important is the need to ensure that the codes are enforced and supervised. Since grading codes were introduced in the 1950s, the strengthening of enforcement and supervision has been as important as the strengthening of the codes themselves in the increasingly successful efforts of local authorities to restrict slope failure. But just as the rigour varies through time, so it varies spatially today, even between cities that have adopted identical codes. Certainly it is now high in cities like Los Angeles and in Los Angeles County, where professional engineering geologists and soils engineers are employed to supervise the planning and development of sloping terrain.

This brief review of the complex building codes in Los Angeles County leads to several conclusions. The first is that there is a widespread recognition and acceptance of the fact that geomorphological hazards on slopes require control through building codes. The second is that the codes vary from the conservative and permissive (as in the CBL, at least until 1962), to the more mandatory and severe (as in CLABC). Controls within the county are generally somewhat more severe than those elsewhere in southern California (Waldrep 1966), perhaps because the Los Angeles experience is greater. Third, controls in different areas have tended to become more severe and more rigorously enforced, and also more similar, as experience of the problems has increased, especially in those cities most directly affected. Fourth, administrative agencies often reserve the right to impose more rigorous controls or to relax restrictions as appropriate; in general, the codes are flexible. Finally, the history of the codes reveals the growing recognition that successful slope development depends on the integration of the efforts of different specialists. This discussion has focused on the geomorphological contribution that is associated mainly with engineering geologists and soils engineers; but others with important and formally recognised rôles include developers, architects, planners and contractors.

Spatial and temporal variability in the severity of building and zoning codes and in the rigour of their implementation are partly responsible for the fact that the codes are not completely effective in preventing slope failure, and they help to explain the pattern of storm damage to slopes. But the generalised controls in codes do not, and probably cannot, apply effectively to the infinity of special local circumstances without being so restrictive that all hillside development would become prohibitively expensive. Furthermore, controls through codes can only be as enlightened as scientific and technological understanding allows, and this is constantly changing. And ultimately, the competence, motivation and resources of those involved will determine the success of a slope-development enterprise. Not least amongst those involved is the multitude of individual property owners whose fundamental rôle in property maintenance is largely uncontrollable.

CONTROLLING SEDIMENT AND FLOODS

Plans and progress

INTRODUCTION. The improvements in monitoring and predicting water and sediment movement in Los Angeles County described above, and progress in the perception of the problems, have accompanied a series of plans during the past 70 years that culminated in the effective protection of most urbanised areas of the county that are potentially vulnerable to flooding. The history of these developments has been admirably chronicled by Bigger (1959). Here, attention will be focused on the nature of the solutions and progress towards achieving them. Figure 3.4 summarises the main relationships among the geomorphological system (see Fig. 1.1), the principal management agencies, and the main methods of control.

Efforts at controlling water and sediment movement focus on hydrological and geomorphological principles and empirical evidence derived from the experience of flood events. Fundamentally, three approaches have been adopted: to preclude development in potentially hazardous zones; to abate the problems by limiting the production of sediment and water from slopes; and to accommodate the supply of water and sediment in suitably sized receptacles. In the discussion of the water and sediment problems throughout this study, emphasis has been placed on geomorphology. In the context of discussing controls, it is here essential to emphasise another fundamental consideration underlying decision making. The Los Angeles area was for many years short of water, and the occasional floods were seen (and still are seen) as a valuable resource as well as a potential threat. Thus, water conservation figures prominently as an objective within flood control programmes (Bigger 1959,

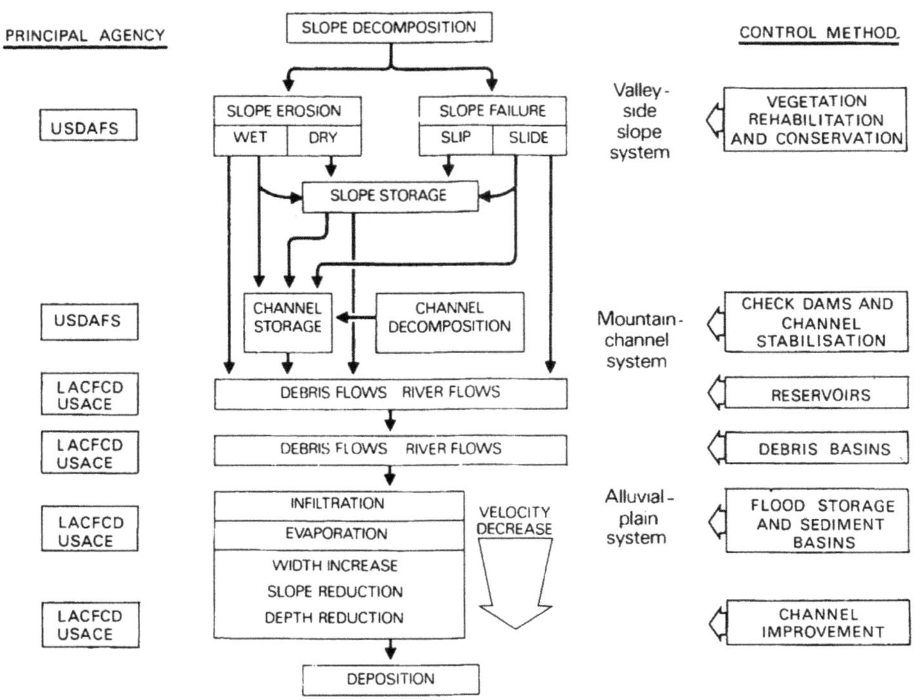

Figure 3.4 Simplified relations among the geomorphological system (see Fig. 1.1), the principal management agencies and the main methods of sediment and flood control.

LACFCD 1961, Turhollow 1975). It is embodied in the Los Angeles County Flood Control Act (section 2), whose aim is to 'provide for the control and conservation of the flood, storm and other waste waters . . .'

The major methods of abating or accommodating the supply of water and sediment were well known in principle by the time the LACFCD was established in 1915 (see Tables 2.2 and 3.1). Abatement controls relate essentially to rehabilitation and conservation of watershed vegetation. Accommodation controls fall into either 'upstream' or 'downstream' categories: the former include check dams and other channel-stabilisation structures, and mountain-canyon reservoirs; downstream, debris basins, water-spreading reservoirs and channelisation methods are pre-eminent.

What has changed since the creation of the LACFCD is not so much the types of control measures available, but the improvement of their cost-effectiveness and design, their integration into increasingly comprehensive management plans, and the partition of responsibilities for them among different agencies.

A SEQUENCE OF PLANS. The major stages in the development of a comprehensive flood-control policy are summarised in Table 3.6. Before 1914, flood protection was a local problem that became loosely organised into seven protection districts. The trauma of 1914 led to the creation of the integrated Los Angeles County Flood Control District, with the passing of the Los

Table 3.6 Major stages in the development of plans for debris and flood control in Los Angeles County.

1915, 12 June	Los Angeles County Flood Control Act
1931+	Formulation of comprehensive flood-control plans by LACFCD
1934, 26 September	Presentation of LACFCD plan to county
1936, 22 June	Federal Flood Control Act (PL 74–738)
1938–41	Preparation of plans by USACE, USDA and LACFCD

Los Angeles River Flood Prevention Project (watershed areas)		*Los Angeles County Drainage Area Project* (LACDA)	
1941	Memorandum of Understanding (USDA 1941) and congressional authorisation	1941	Establishment of LACDA (Flood Control Act PL 77–377)
		1945	State of California Water Resources Act
1954	The Watershed Protection and Flood Prevention Act (PL 83–566)		
1959	LACFCD 'Blue Book' report		
1961	USDAFS review report ('Green Book')		
1974	USDAFS review report		
		1975	LACDA review initiated

Angeles County Flood Control Act on 12 June 1915. This important administrative area did not include the whole county (see Fig. 2.1), because of the difficulty of securing the two-thirds majority vote required for the insurance of bonds to support such a county-wide authority; only a simple majority is required for 'district' bonds (Bigger 1959). This decision was to have far-reaching consequences later in the century, when urban growth extended into the northern areas of the county beyond the district's boundaries. It is in these areas that some of the most serious flood hazards are found today.

The LACFCD's first flood-control proposals (see Table 2.2), modest but relatively comprehensive in terms of the need for protection at the time, were modified by Reagan (1917). They were implemented and extended in the relatively dry years of the 1920s, and funded largely by bond issues, local taxes and state aid. By 1931, up to a third of mountain watersheds were believed to be 'controlled' by 12 regulatory dams and check dams in mountain canyons; 11% of channels were controlled by permanent works, and a further 24% by temporary works, and 550 ha of spreading grounds were established (Eaton 1931).

Nevertheless, it had become clear by the late 1920s that the plans of 1915–17 and the progress in implementing them were inadequate to meet the needs of rapid urban growth, and a new, comprehensive, plan was required. The foundations of this plan were laid by the district (Eaton 1931). Provisions were made in 1931 for new control measures including dams, spreading grounds, channel regulation and debris basins, for reserving rights of way along channels, and for reserving reservoir sites before urban developers could engulf them; 'immediately needed' projects were estimated to require $33 million. Another plan was submitted to the county by the LACFCD on 26 September 1934 (LACFCD 1934), which included details of proposed and urgently needed channel engineering works, debris basins, and regulating reservoirs.

By 1935, the problems of flooding were attracting more federal interest. The first Flood Control Act (PL74–738), passed in 1936, established federal involvement in flood-control projects, mainly through the USDA (Forest Service and Soil Conservation Service) and the USACE – the former having responsibility primarily for headwater areas, the latter primarily for major construction projects. Both agencies had built up experience in the region during the 20th century, but not mainly in the context of river flooding. After 1936, plans for flood and sediment control were developed by both agencies in collaboration with the LACFCD, the latter holding the central rôle of integrating policies, and also having the tasks of managing existing facilities, constructing minor works, building the storm-drain system (mainly since 1952), and providing advice on flood-plain zoning to the county and other local authorities.

Efforts at developing an 'upstream' programme (the Los Angeles River

Flood Prevention Project) were formally incorporated into a memorandum of understanding among the Secretary of Agriculture, Los Angeles County, the LACFCD, the City of Los Angeles and a number of other cities (USDA 1941, House of Representatives 1941, 77th Congress, Document *426*). The initial proposals related only to the Los Angeles River Watershed, but they were extended to include the San Gabriel and Santa Clara drainage areas in the Watershed Protection and Flood Control Act of 1956. The memorandum was based on extensive prior studies generated mainly by the USDA Field Flood Control Coordinating Committee No. 18 (USDA 1938, 1949, USDAFS 1939, Conservation Association of Los Angeles County 1938). The proposals (summarised in Table 3.7) had the major objectives of increasing the life of downstream structures, reducing and controlling peak flood flows, reducing flood damage and deposition, preventing channel clogging during floods, improving recreational resources of watersheds, and improving yields from farmlands (USDA 1940). Water conservation and reduction of property loss by fire would be additional benefits.

The proposals emphasised the need for fire control (construction of fire-truck roads, firebreaks, purchase of water trucks etc. to reduce fire hazard and subsequent erosion), for road improvement (construction of barriers, revetments, deflectors etc. to reduce erosion), for channel protection (especially through check dams), and the rôle of farmland improvement (including terracing, grazing control and other measures to reduce erosion and increase production). The USDA was to implement the plans mainly through the soil conservation programme on agricultural land, and the forest service programme in the watershed areas. Funds for the work – originally estimated at \$11 416 169 – were to come mainly from federal sources (73%), but also involved local contributions (e.g. 50% of funding on non-federal lands and contributions towards costs of rights of way). The plan was due for completion by 1954; but by 1972 the fire control programme was 80% finished, the cover improvement phase was 30% accomplished, road restoration was 71% accomplished, and channel improvement only 29% complete (Fig. 3.5, USDAFS 1974). By 1982, there were some 357 km of fire roads and trails, approximately 394 km of fuelbreaks, together with numerous fire-fighting services (USDAFS 1982). The fire-control programme protected an estimated 630 km^2. The channel-improvement programme had installed over 400 channel-stabilising structures or check dams and four major debris basins, which together had trapped some 2×10^6 m^3 of debris. The critical area stabilisation programme had accomplished over 22 000 ha of seeding and planting for slope stabilisation, 317 km of road stabilisation, 90 km of channel stabilisation, and some 48 km of gully control (USDAFS 1982). During the period of implementation, changes were made to the plans, most notably in the programme for check dams and channel stabilisation, in the approaches to debris disposal (USDAFS, 1974), and in the adoption of 'prescribed burning techniques' (see p. 34, USDAFS 1982). Other changes were more political:

Table 3.7 Improvement programme for the Los Angeles River Watershed: estimated installation costs of measures in aid of flood control (USDA 1941).

	Secretary of Agriculture	State of California	Los Angeles County	Los Angeles City	Glendale City	Ventura County	Farmers and other individuals
fire control	3 197 751[a]	0	112 233	140 337	0	0	0
cover improvements	74 420	0	0	0	0	0	0
mountain road improvement	796 844[a]	147 300	97 455	0	2 000	1 000	31 701
mountain channel improvement (including check dams)	2 868 555[a]	0	996 445	0	0	0	0
farm land improvement	1 054 060	0	0	0	0	0	1 120 058
debris basins and channels	388 005[a]	0	388 005	0	0	0	0
Total ($)	8 379 635	147 300	1 594 138	140 337	2 000	1 000	1 151 759

[a]Includes 50% of the total estimated installation costs of measures on non-federal lands, exclusive of right-of-way costs.

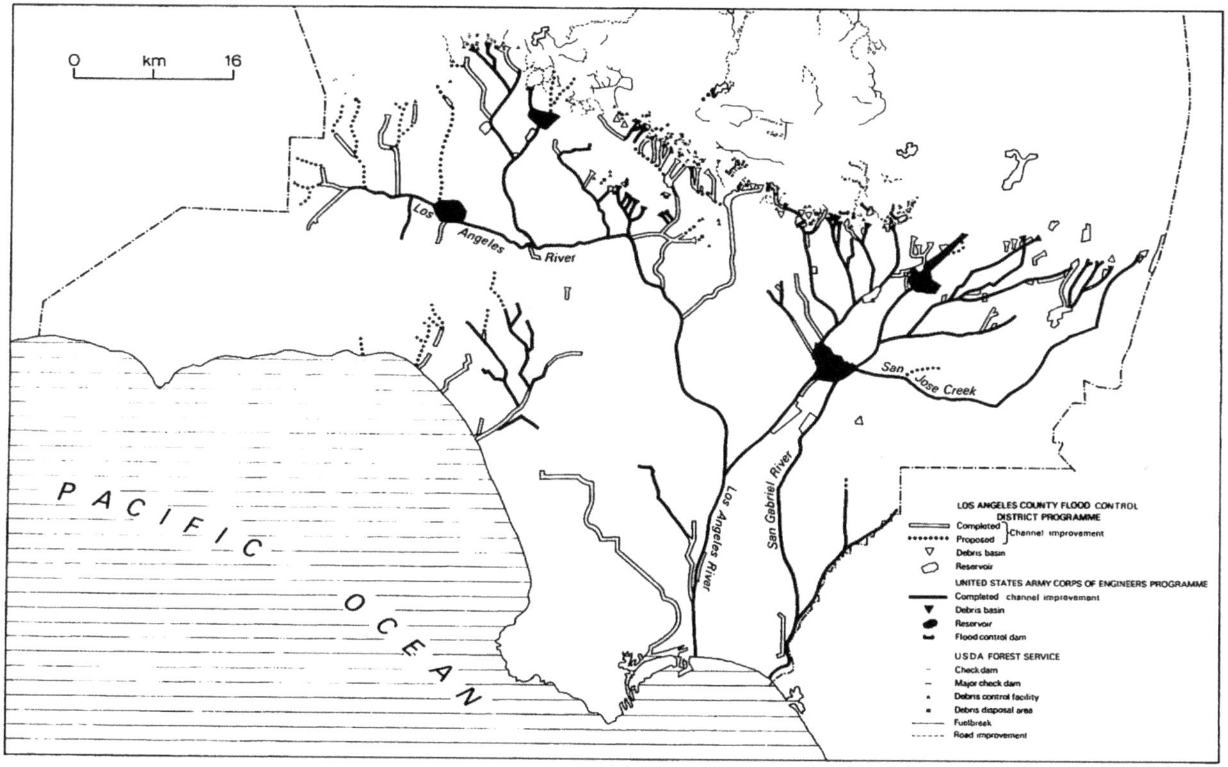

Figure 3.5 The main achievements of flood-control plans in Los Angeles County (after USACE 1978b, USDAFS 1982).

in 1981, for example, the proposed budget and the project life were halved, and further cuts were made in 1982.

The USACE and the LACFCD began to formulate a common approach to flood control in the late 1930s, building on the comprehensive plan of 1931–34, and directing attention towards the objectives of establishing basins to collect debris at the mouths of canyons, of building flood-control basins to contain peak flows and regulate downstream flows, and of rectifying and stabilising natural channels across the coastal plain (USACE 1938, Turhollow 1975). The 1936 Flood Control Act (PL74–738, House of Representatives 1936, Document *455*) authorised the construction of reservoirs and principal flood channels 'in the Los Angeles and San Gabriel rivers and Ballona Creek and tributaries thereof' at an estimated construction cost of $70 million, plus $5 million for lands and damages. Work began at once, the employees of the USACE rising from a handful to over 16 400 in 1936 – a fact that helped to relieve the pressure of unemployment (USACE 1938). The 1938 amendment to the act made provision for federal purchase of title to lands, easements and rights of way. The proposed general plan (Table 3.8, US House of Representatives, 1940, 76th Congress, Document *838*) extended the original investment from $82 million to over $286 million for construction, and involved annual carrying charges of $13.5 million. From 1936 to 1938, the USACE prepared and submitted numerous proposals for the control of the Rio Hondo, and the San Gabriel and Los Angeles rivers, and these were ultimately incorporated into the Los Angeles County Drainage Area project (LACDA) as approved in the 1941 Flood Control Act (USACE 1938, US House of Representatives 1940, Document *838*). It was estimated that the potential benefits from the proposed improvements (including prevention of flooding, reduction of repair, replacement and cleaning costs, reduction in unrecoverable income losses, less interruption to services, and water conservation benefits), spread over a 50-year period, would amount to about $20 million p.a. The overall estimated benefit–cost ratio was 1.52 : 1 (with individual items ranging from 4.80 : 1 to 1.02 : 1).

Successive flood-control acts modified plans or the rôle of the federal agencies. For example, the 1944 act altered the basis of financing by permitting the USACE to construct and maintain public park and recreation facilities at its sites; the 1948 act allowed the USACE to undertake certain small flood-control projects and emergency work without time-consuming prior congressional approval. The State's Water Resources Act of 1945 was another important step in the state's financial support for flood-control measures, providing assistance for buying lands, easements, rights of way etc.

From 1936 onwards, the USACE, working with the LACFCD, produced many surveys, plans, damage reports and other documents relating to the design, construction, operation and cost-effectiveness of structures proposed in the LACDA plan. By 1971, the plan was almost complete (Table 3.9, Fig. 3.5). All the main flood- regulating dams, debris basins and spreading grounds

Table 3.8 Principal components of the LACDA plan (US House of Representatives 1940, 76th Congress, Document *838*).

Project item	Estimated costs
Los Angeles River Basin	
pre-1940	
Hansen Flood Control Basin	10 380 000
Sepulveda Flood Control Basin	6 821 000
River channels (including debris basins)	50 082 573
1940	
Lopez Flood Control Basin	4 779 000
Channel improvements	67 063 000
	139 125 573
San Gabriel Basin	
pre-1940	
Santa Fé Flood Control Basin	10 017 000
River channels (including debris basins)	2 086 000
1940	
Whittier Narrows Flood Control Basin	16 983 000
Channel control and improvements	57 142 300
	86 228 300
Rio Hondo Basin	
pre-1940 } channel work	1 205 613
1940 } channel work	21 378 600
	22 584 213
Ballona Creek	
pre-1940	1 950 100
1940	18 276 500
	20 226 600
Total	268 164 686

were in operation, and most channel improvements were finished. The main area still requiring attention was the Santa Clara drainage, which had become a new focus of urban development. The LACDA programme had involved the USACE spending $332 million (by 1969), and the LACFCD spending $90 million (by 1971) in construction projects and acquiring rights of way etc. (USDAFS 1974). In addition, 'supplemental flood-control improvements' in the LACDA area had cost local interests some $765 million by 1965 (Turhollow 1975). A major comprehensive review of LACDA has been in progress since 1975 (USACE 1982).

Avoidance measures Amongst the methods of flood adjustment that do not require engineering work, the control of land use in flood-hazard zones, and flood-hazard avoidance are pre-eminent. Control is mainly through zoning and subdivision ordinances, most of which recognise the possibility of

Table 3.9 Aims and achievements of the LACDA Comprehensive Plan (USDAFS 1974).

	Proposal	Achieved 1971
flood-regulating dams	20	20
debris basins	105	76
spreading grounds	30	27
permanent flood channels (km)		
Los Angeles River	77.2	77.2
San Gabriel River	46.7	46.7
Rio Hondo	20.9	20.9
Ballona Creek	16.1	16.1
Laguna-Dominguez	27.4	27.4
Santa Clara	54.7	0.0
tributary channels	790.2	519.8
	1033.2	708.1
storm drains (km)	3329	2399

flooding. The county zoning provisions are fairly typical, prohibiting the construction in zones liable to flooding of buildings 'designed for living purposes, as a hotel, lodging house, school or institution or home for the treatment of convalescent persons, children, aged persons, the wounded or mentally infirm . . .' (Article 4, Chapter 3 of the Los Angeles County Zoning Ordinance). The county subdivision ordinance (Section 158) reserves the right to disapprove, or require protective improvements to, developments on land subject to flooding or inundation. Hazard avoidance is not only often cheaper and generally preferable to engineering control of flooding, it may also bring with it land-use benefits, such as the provision of open space and recreational areas. At the same time, flood-plain development regulations must be effectively implemented if they are to be successful, and this has not always been the case. Leopold and Maddock (1954, p. 21) suggested that in the past:

> The pressure for expansion of a rapidly growing city has been so great that ordinances and regulations either have been ignored or else only a minimum amount of flood control has been provided concurrently with development. An inevitable consequence has been that expenditures for flood control per square mile of drainage basin, by federal and local agencies and by individuals, are greater in the Los Angeles area than anywhere else in the United States.

Critical to the successful avoidance of flood hazard by zoning and similar management activities is the knowledge of flood-hazard areas and, equally important, flood-protected areas available to land-use planners. The delimitation of hazard and protected zones is thus a first and fundamental requirement. It is provided for the county and most other local authorities by the LACFCD, by the federal insurance rate maps and surveys (which have superseded the *flood-plain information reports* of the USACE), and by such other reports as those for the state as a whole by the Natural Resources Agency (State of California, The Resources Agency 1978). The LACFCD is central to this work. It provides, *inter alia*, a manual (LACFCD, no date) which advises engineers on how to identify flood and debris hazards and how to design protection against them. In unincorporated county land, the LACFCD's approval of flood-control plans is normally required before the Regional Planning Commission will accept development proposals; in incorporated cities, the LACFCD's rôle is advisory. The LACFCD distinguishes, in this context, between flood zones where flow velocities are below 3.05 m/s and where, under certain conditions, low-density development may be permitted; and zones where velocities are higher, and flood-control improvements are required (LACFCD 1981). Unless the LACFCD has already delineated areas reserved for flood flows, the responsibility for delineation (like the subsequent engineering design) rests with the developer.

Methods of abatement and accommodation

'UPSTREAM' EMERGENCY MEASURES. Emergency measures designed to control soil erosion from surfaces freshly laid bare by construction work or fire have become widely adopted throughout the county. These measures mainly involve emergency sediment-accommodation structures such as desilting basins, soil conservation practices on agricultural land, or re-vegetation. Such measures are incorporated, for example, in the City of Los Angeles' *Emergency and disaster manual* (no date), and in its building codes. There are detailed procedures for adopting, funding, implementing and monitoring these activities (e.g. City of Los Angeles, Department of Public Works, Special areas S004–0269, February 5, 1969). In the case of emergency measures following a fire, the procedures require collaboration between city and county agencies, and maps have to be prepared showing the burned area and the location of proposed soil-accommodation structures. Emergency reports, which may suggest erosion-control practices following fires, are also commonly prepared (e.g. Los Angeles County, Department of County Engineer 1973). Soil conservation on agricultural land may be informal, or it may be organised (in collaboration with the USDA) through soil conservation districts, of which there are four in Los Angeles County.

Emergency vegetation rehabilitation, especially following fires, commonly takes the form of broadcast re-seeding with fast-growing grasses (sometimes using helicopters and fixed-wing aircraft), or row plantings of barley (Corbett & Green 1965). Emergency revegetation of private land is the responsibility of the California Division of Forestry (Phillips 1971); revegetation of federally owned land in Los Angeles County is in the hands of the USDAFS and the Bureau of Land Management; the Los Angeles County Fire Department is also involved. Inter-agency collaboration is essential – as after the fires of 1970 when some 44 000 ha were re-seeded in Los Angeles County (Phillips 1971). The benefits of such action are assumed to lie in the surface protection provided by relatively quick-growing grasses and the consequent reduction of erosion (Corbett & Green 1965). One problem of emergency seeding is that it is not appropriate everywhere, for the soils must be suitable. Others are that early rains may cause erosion before the seeds germinate, or winds may blow the seeds away. Therefore, one further emergency approach is to protect the surface during the critical germination period by spraying the ground with chemical binders such as asphalt emulsion. Early tests of binders were not very encouraging (Krammes & Hellmers 1963). One bonus from emergency seeding is the replacement of chaparral by grass – a change that may have the effect of reducing soil-moisture losses, promoting percolation to groundwater, and thereby increasing water yield (Crouse 1961, Hill 1963).

LONGER-TERM 'UPSTREAM' MEASURES. Long-term vegetation control in watershed areas has been a matter of controversy since it was first discussed (Los Angeles County Board of Supervisors 1915), mainly because of the different objectives of vegetation policies and the changing views of the nature of chaparral vegetation. In the conservation debate of the 1930s, for example, the relative merits of creating tree or brush covers in the mountains and their respective rôles in reducing erosion were disputed (Conservation Association of Los Angeles 1932). Fire exclusion policies prevailed for many years, but these have been modified recently by the recognition that the dry fuel in a stand increases with time, making older stands more combustible, that fire is fundamental to chaparral ecology, and that fire control may be more effective through a policy of scheduled burning and fire containment (Hanes 1971, Countryman 1974). Moreover, policies of fire control are not necessarily those that will reduce slope erosion – except in so far as they may reduce the magnitude of fire-induced accelerated erosion in burned areas – because such policies often involve the creation of vegetation-free, erosion-vulnerable surfaces, as along fire tracks, firebreaks and roads in the mountains, and around buildings. As discussed above (see p. 14), conversion to grass may actually increase the potential for soil slips under certain circumstances; and the impact of controlled burning on sediment loss is uncertain.

The effectiveness of various vegetative methods in controlling erosion following a fire was often strongly affirmed (LACFCD 1931), but experimental support for it is quite limited. A study by Rice *et al.* (1963) in the San Dimas Experimental Forest is particularly instructive. In this study, 20 similar burned watersheds were selected for treatments, and water and sediment yields were monitored from each. The treatments included sowing with annual grass mixtures or perennial grasses, and a selection of mechanical erosion-control measures such as side-slope stabilisation by barley planting along the contour, contour trenching, and channel stabilisation with check dams. The conclusions included the views that perennial grass seeding was relatively unsuccessful, annual grass seeding can be justified as an emergency erosion-control measure but was relatively unsuccessful in the first year (when its effects may be most needed), and side-slope stabilising by contour furrow planting was the most effective method used.

The long-term control of sediment yield in upstream areas rests in part on three other strategies. The first requires the control of 'urban facilities' (houses, roads etc.), either through county or city ordinances, or through the proper design of facilities on public land. The second involves incorporating appropriate drainage controls into the design of mountain roads (e.g. State of California, The Resources Agency 1971).

A third strategy of sediment control in the mountains focuses on the construction of debris-retaining structures, which include crib structures, check dams, and reservoirs. Crib structures and check dams are designed to trap sediment in channels and, in so doing, to assist in stabilising both channel banks and the vulnerable lower valley-side slopes by reducing gradients, increasing valley-floor width, and thus locally reducing flow velocity, promoting the growth of protective vegetation, and retarding channel erosion and movement of large boulders; in addition, they increase water storage and promote water conservation (Eaton 1931, Ferrell 1959, Ruby 1976). The remedy of constructing small barriers across channels is an old one, drawn originally from experience in the quite different Swiss environment (Los Angeles County Board of Supervisors 1915). Between 1914 and 1934, it is variously estimated that 1000–4000 check dams were constructed to help store debris and stabilise channels (LACFCD 1959, USDAFS 1973b). Two reports (Skafte 1930, Eaton & Gillelen 1931) assessed these early, relatively flimsy barriers which normally comprised only loose blocks or rock-and-wire structures and rarely survived major runoff events. Both reports alluded to the inadequacies of check dams, but on balance favoured them as a means of sediment control. Some of these early structures were destroyed in the storms of 1934 and, indeed, their sudden failure may have contributed locally to debris pulses in flows.

But the philosophy of check dams was more persistent than the early structures. They continued to be advocated for small-scale upstream sediment and channel control, principally by the USDA, and a proposal to construct 852 check dams was incorporated in the Los Angeles River Flood Protection Project, authorised by Congress in 1941 (House of Representatives 1941, 77th Congress, Document *426*).

Designs were improved. Sturdier structures than those that failed in 1934 were introduced in Brand Canyon (near Glendale) in 1938, including 12 mortar-rubble masonry-arch dams, 5 mortar-rubble masonry-granite dams and many loose-rock dams. Various construction techniques were used during the development of check dams in the Arroyo Seco Flood Control Project in the 1940s; and the USDAFS and the LACFCD combined to build concrete crib structures in Cooks Canyon in 1956 (LACFCD 1959). Designs are still being improved (Li *et al.* 1979).

The location of structures is strongly influenced by channel slope, because costs of stabilisation increase directly with slope (LACFCD 1958). Normally, channels with slopes over 20% are not stabilised, and the lower the channel slope, the greater will be the length of channel stabilised.

The programme was given a strong fillip by three reports. The first was on channel stabilisation feasibility in the San Gabriel and Verdugo mountains (LACFCD 1958), the second was an assessment of the Los Angeles River Flood Protection Project (USDAFS 1961), and the third was on debris reduction studies (LACFCD 1959). The last of these, for example, not only recognised an apparently successful debris-accommodation rôle for such structures, but also emphasised as being relatively more important the rôle of the debris wedges they created in stabilising channels and slopes. The report proposed 35 promising watersheds for future construction. By 1972, over 300 of the 852 structures approved in 1941 had been built. Most survived the 1969 storms, but some were breached, often by large boulders (Ruby 1976).

The success of this programme is a matter of debate. Clearly, check dams trap sediment, and they actually trapped more than predicted in the early designs, mainly because of the mounds which develop on the sediment cones adjacent to them. These mounds provide a temporary increase in storage that is usually flushed out in major floods (Ruby 1976). But the dams are only effective in reducing sediment yield for relatively short periods. Some of them were filled to about 60% of design capacity immediately following construction (USDAFS 1973b), and some only reduced yield for about seven years (Ruby 1976).

The argument that sediment wedges behind dams serve to stabilise valley-side slopes and reduce side-slope erosion is important. The LACFCD (1958) comparison of Brand (controlled) and Sunset (uncontrolled) canyons showed, for example, that the area of channel and adjacent side-slope erosion was only some 56 m^2 per 100 m in the former, and 174 m^2 per 100 m in the latter. But the evidence is not entirely convincing. Ruby (1976) suggested, for example, that this effect might be reversed if a major flood lowered the sediment cone and thus exposed the previously stable valley-slope deposits to erosion. It is also possible that if flow is divided by the central sediment mound, erosion may actually be concentrated along valley sides. Nevertheless the sediment wedges certainly broaden the channels, lower channel gradients locally, reduce downcutting, promote sediment-retarding vegetation growth, reduce velocities of flow, and encourage water infiltration. But the effect of these changes on sediment yield, and the duration of the effect if it exists, are probably not great. Ruby (USDAFS 1973b) concluded that there was a slight permanent reduction due to these causes.

From 1955 to 1972, $7.9 million was spent on check dams (USDAFS, 1973b). Ruby (1976), in an analysis of the cost-effectiveness of this investment, concluded that some dams are more effective than others, but that, overall, the cost-effectiveness is marginal or negative. Nevertheless, he showed that check dams did perform (at least temporarily) the functions required of them in the 1941 plan: they have retarded channel downcutting and debris movement, raised and widened channel beds, reduced channel velocity, reduced the movement of large boulders, and controlled the direction of flows.

Reservoirs in the mountains are a final and very significant means of trapping sediment in 'upstream' areas, as well as on the plains (see below).

DEBRIS BASINS. The most important 'downstream' response to the debris problem has been the construction, over a period of some 50 years, of debris basins designed to catch coarser sediments at or near the apices of alluvial fans. A total of 112 debris basins were constructed in the LACDA area by 1979. As Eaton remarked, 'Their physical locations are closely fixed, between the canyon mouths and the deltas, at points sufficiently towards the canyon mouths to prevent channel-cutting around them, and as far down on the deltas as feasible toward the flatter slopes where excavation costs will be least' (Eaton 1936, p. 1324). The structures are designed to accommodate large influxes of sediment. For example, Eaton recommended a design capacity of 22 132 m^3/km^2 to 44 264 m^3/km^2 (Eaton 1936). Tatum (1963) reported that design capacity was originally based on a flat rate of between 14 754 and 29 509 m^3/km^2, and that later it was based on an enveloping curve of debris inflows plus 100%. Since 1959, their design capacity has been based on the debris-production curves developed in the LACFCD's (1959) *Debris reduction studies*, and on the assumption that the debris slope will be 60% of the original channel bed (LACFCD 1965). A design capacity of 59 000 m^3/km^2 is usually adequate because this is approximately the maximum recorded sediment yield. Tatum's (1963) analysis led him to recommend a design quantity equivalent to 64 920 m^3/km^2 – a figure somewhat higher than that commonly used.

Unlike the crib structures, debris basins are re-usable controls. They are usually cleaned out when their capacity is reduced by 25% or more, or after major storm events. The debris (which excludes the fine sediment that passes through the basin outlet towers) is expensive to remove. While some of it can be used as fill in grading work and other construction activities, there are growing problems of finding suitable disposal areas for the surplus. Disposal in small canyons is hazardous and has been stopped; old quarries are desirable but not common. Sites are becoming more distant from the sources of debris, and more expensive. As a result, costs are rising. Thus, for example, the average cost of removing a cubic metre of debris from debris basins rose from $1.85 in 1965–66 to $3.88 in 1975–76 (USDAFS 1979c) – a real increase when inflation is taken into account of 22.5%. Another serious problem is that some basins have been overtopped during storms, leading to downstream damage, as in Shields Canyon in 1978. But in general the basins have been very successful in restraining the movement of debris from small catchments on to vulnerable alluvial fans.

In addition to the major debris basins built by public agencies, smaller debris dams may be designed, built and cleared out by private developers to protect subdivisions and to meet the LACFCD requirements (LACFCD 1965). Sometimes these may be temporary features. Private debris-control structures have become more important in recent years.

MAJOR DAMS, RESERVOIRS AND CHANNEL IMPROVEMENTS. These constitute the greatest investment in flood and sediment control. There is an important distinction between dams in the mountains, and the major flood-control basins and spreading grounds on the plains. All act as controls on flooding and trap sediment, but the latter are also concerned with water conservation. The comprehensive LACDA plan involved the construction of 20 flood-regulating dams, and by 1974 these had all been completed. The dams, together with other sediment- and flood-retaining structures, have undoubtedly been effective in reducing flood hazard in the areas downstream of them. As a result of a net accumulation of sediment in the reservoirs behind the dams, their storage capacity is declining (Kenyon & Coakley 1960). Bruington (1982) cited the example of the reservoir behind the San Gabriel Dam which has a tributary area of 416.4 km^2. Its original capacity was 65.37×10^6 m^3 in 1938. By 1980, it had received 32.8×10^6 m^3 of sediment. Of this total, 10.6×10^6 m^3 remains in the reservoir (reducing its capacity by 16%), and the remainder (22.1×10^6 m^3) has been removed by sluicing, or on debris clearout contracts issued by the LACFCD at the enormous cost of $20 million (1981 prices).

The final major method of flood and sediment control in 'downstream' areas has been to improve the drainage channels permanently, and to provide an efficient storm drain network in developed areas. Channel improvement involves a range of strategies including the building of permanent lined channels of dimensions capable of carrying predicted floods, and the construction of levées, bridges and other features. These improvements do more than provide a permanent channel system capable of efficiently removing the runoff from storms; they also remove the indeterminacy of flow direction that characterises alluvial fans, thus removing potential flood risk from large areas. Over 1031 km of channel improvements were planned; by 1971, over 700 km had been completed. The storm drain system is a major feature of LACFCD work in the county (LACFCD 1970b). The agency not only builds and maintains its own structures, but it also evaluates and approves drains proposed by developers. Design criteria for such work are well established (LACFCD 1965), and include details of such features as manholes and debris-control structures. Over 3329 km of storm drains were planned; by 1971, over 2400 km had been completed.

4

CODA

Costs and benefits in context

Within Los Angeles County

It is important to answer the question, 'Is damage caused by geomorphological hazards in Los Angeles County declining over time as a result of management efforts?' Unfortunately, there is not a simple answer, for several reasons.

In the first place, the damage data for storms (such as those given in Part 2) are far from perfect. They do not usually discriminate between damage related to geomorphological hazards and that arising from other causes. And, although the damage estimates have become increasingly reliable over time – especially since the adoption of USACE standards (USACE 1943) – the figures are certainly underestimates of true storm-hazard costs. Reasons for this include the fact that damage data are not comprehensive, that the emphasis is on physical damage and intangible losses are relatively neglected, and the difficulty of determining slope-damage costs before 1969.

Another problem is that damage data need to be adjusted for inflation. The consumer price index provides one adjustment criterion (Fig. 4.1, curve A; US Department of Commerce 1975). The best estimates of damage are shown as a cumulative curve adjusted for this index in Figure 4.1, curve Bi. An alternative, and arguably more realistic, index relates to the rise in construction costs, such as that of the *Engineering News – Record*, because damage repair costs more closely reflect construction costs. This index shows a higher rate of rise than the consumer price index. Damage data adjusted for it are shown in Figure 4.1, curve Bii, and in Table 4.1.

A third problem is that the inflation-adjusted damage data need to be viewed in terms of both the population of the region and the available wealth of the

Figure 4.1 Selected damage and cost figures, Los Angeles County. (A) Dollar inflation (1967 = 100) (from US Department of Commerce 1975). (B) Cumulative curves of storm damage from major storms only; based on (i) present study, best estimates, corrected for inflation using the consumer price index (1967 = 100), and (ii) USDAFS (1974) estimates corrected to 1973 base, using the *Engineering News – Record* construction cost index. (C) Cumulative curves of deaths from storm hazards in Los Angeles County. (D) Estimated costs of debris disposal, cumulative curve, 1974 prices, for the Los Angeles River Watershed only (USDAFS 1974).

DOI: 10.4324/9781003168645-13

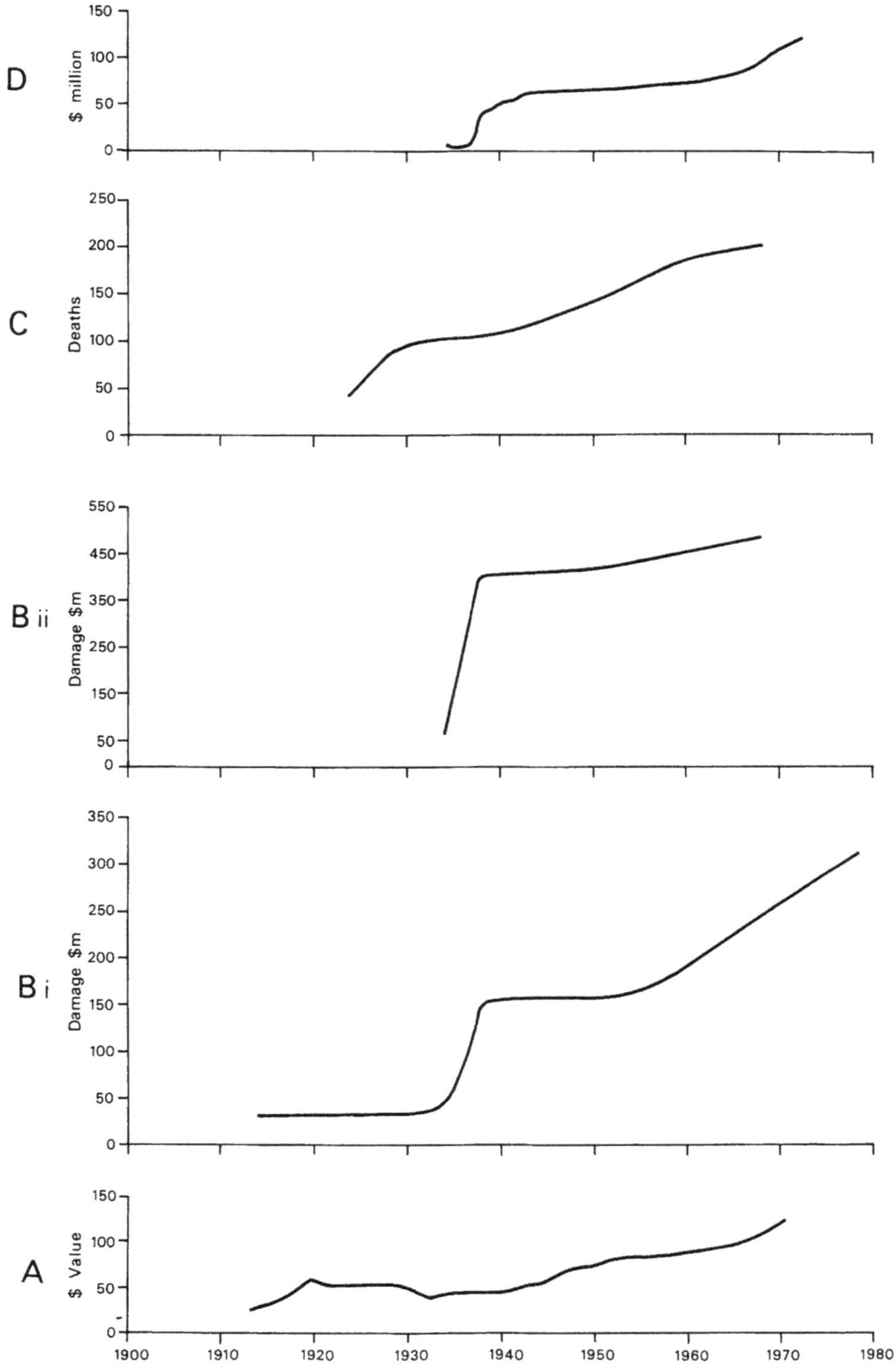

D
$ million
150
100
50
0
C
Deaths
250
200
150
100
50
0
B ii
Damage $m
550
450
350
250
150
50
0
B i
Damage $m
350
300
250
200
150
100
50
0
A
$ Value
150
100
50
0
1900
1910
1920
1930
1940
1950
1960
1970
1980

Table 4.1 Data for indices of storm damage, Los Angeles County.

Year	Damage, D (\$m)	Damage, D^a (\$m) Consumer price index (1967 = 100)	Damage, D^b (\$m) Construction cost index[c] (1967 = 100)	Population P (m)	GNP per capita index[e] (1967 = 100)	Precipitation seasonal total (cm) downtown Los Angeles[f]	Precipitation storm total (cm) downtown Los Angeles	$\frac{D^a}{P \times \text{GNP per capita}}$	$\frac{D^b}{P \times \text{GNP per capita}}$	$\frac{\text{Deaths}}{P}$ (%)
1914	10.0	33.22	114.9	0.7[d]	37.28	60.07	17.88[g]	1.27	4.40	—
1938	45.0	106.63	205.54	2.45[d]	43.67	59.51	28.09[g]	0.99	1.92	0.00204
1969	97.0	88.34	81.71	3.0[d]	105.35	58.34	43.68[g]	0.119	0.11	0.00105
1978	94.0	48.10	36.60	7.03[d]	127.66	69.77	48.33[c]	0.053	0.0407	0.00022

[a]Best estimate of total damage, in original \$ values. Slope-failure damage estimates included for 1969 and 1978 are for private property only.
[b]Index source: US Department of Commerce, Bureau of the Census (1975), and annual reports thereafter.
[c]Index source: US Department of Commerce, Bureau of the Census (1976), and *Engineering News – Record*.
[d]US Department of Commerce, Bureau of the Census data, 1978 is an estimate (USDC, Bureau of Census 1979, P–25/818).
[e]Index source, as for note 2, adjusted by author.
[f]Data source: US Weather Bureau.
[g]*Source:* Simpson (1969).

community. The best way of adjusting the data to reflect these variables is debatable. But as an elementary first approximation, the inflation-adjusted damage data for any given year are divided by the product of the population of Los Angeles County and the gross national product per capita for the US for that year adjusted for inflation to a common base (Table 4.1). Thus, a provisional index of storm damage may be defined as follows:

$$\text{storm damage index, } I_t = \frac{D^1 \,[\text{or } D^2]}{P \times \text{GNP per capita}}$$

where

I_t = index at time t;
D^1 = best estimate of total storm damage at time t corrected for inflation by the consumer price index (1967 = 100);
D^2 = best estimate of total storm damage at time t corrected for inflation by the construction cost index of the *Engineering News – Record* (1967 = 100);
P = population of Los Angeles County at time t;
GNP per capita = index of gross national product per capita at time t (1967 = 100).

The storm damage index values obtained in this way can then be related to the magnitude of the responsible storms. The two simple measures of magnitude selected here are total seasonal precipitation and total storm precipitation, as recorded in downtown Los Angeles. The storm damage and storm magnitude data are shown in Table 4.1 for four major county-wide storms discussed in Part 2 of this book. Figure 4.2 shows clearly that the storm damage indices decline over time, supporting the view that management and related community responses have become increasingly effective in ameliorating storm hazard. Similarly, the ratio of storm fatalities to the population of the county has declined (Table 4.1).

It is also pertinent to ask if the investments in hazard defence are cost-effective. To answer this question is a fearsome task, primarily because so many aspects of hazard impact and management are difficult to cost (e.g. the costs of fatalities, mental disturbance and implementing building code practices), so many estimates include elements of wishful thinking, and many cost–benefit analyses are undertaken by agencies seeking to justify their earlier or proposed actions.

Table 4.2 shows that the approximate total investment in flood control up to 1974 in Los Angeles County was \$1491.2 million, with the majority of funds derived locally. Investment in flood protection is normally only sanctioned if it can be shown, in advance, that it will be cost-effective over a given period of time. There are many estimates in Los Angeles County, and many others that

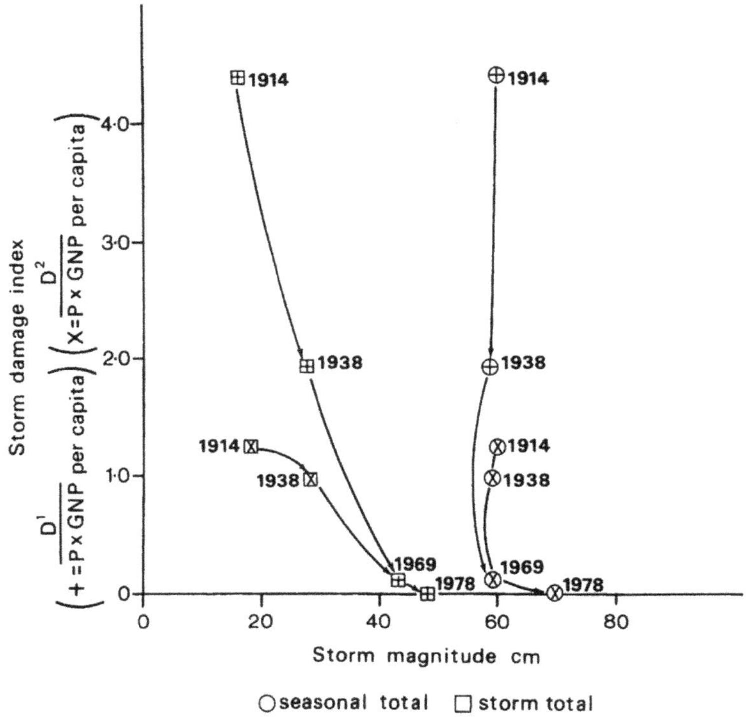

Figure 4.2 Relations among storm damage indices, magnitude and time of occurrence.

demonstrate how, once protection measures are adopted, the investment was worthwhile.

The initial cost–benefit analysis of the LACDA project is summarised in House of Representatives Document *838* (1940). All the proposed developments (it was argued) would show a net benefit over a 50-year period. The estimated benefit : cost ratios of proposed projects ranged from 4.8 : 1 (for the Rubio diversion) to 1.02 : 1 (for Centinela Creek); overall, the ratio was estimated at 1.52 : 1. The assumed benefits related only to flood control, and did not take possible water-conservation benefits into account. The USACE has estimated losses that have been prevented during storms by the LACDA programme. For the 1969 storms, the figure was put at $1013 million (USACE 1969); in 1978, damage prevention was estimated at $1200 million (USACE 1978b). In one sense at least these figures are misleading, because it is surely probable that if the protected land had not been protected, then much of the development subsequent to protection would not have occurred. Nevertheless, given that a substantial proportion of the investment in protection was to defend land already developed, the estimated savings from LACDA projects in 1969 and 1979 together suggest that the investment in flood control was worthwhile.

The original estimates of the Los Angeles River Watershed Project predicted

Table 4.2 Estimated flood-control expenditures of all agencies in coastal Los Angeles County (USDAFS 1974).

Programme project	Expenditure (millions of dollars)
Soil Conservation Service programme, San Fernando Valley	
SCS appropriations to 1972[a]	4.5
LA County Flood Control District to 1971[b]	14.0
Forest Service Programme	
USDAFS Appropriations to 1972[c]	30.5
LA County Flood Control District to 1971[c]	2.5
Los Angeles County Drainage Area Project	
US Army Corps of Engineers first costs to 1968[d]	332.0
LA County Flood Control District to 1971[b]	90.0
Local expenditures for flood control, debris control, and water conservation	
LA County Flood Control District to 1971[e]	995.7
Fire Suppression and Prevention Programme	
LA County Fire Department, fiscal year 68–69 to present[f]	22.0
Total	1491.2

[a]*Source*: Soil Conservation Service, Regional Office, Berkeley, California.
[b]*Source*: *Biennal report for period ended June 30, 1971*, LACFCD
[c]*Source*: Angeles National Forest, Pasadena.
[d]*Source*: *Water resource development by the US Army Corps of Engineers in California*, South Pacific Division, Corps of Engineers January 1969.
[e]*Source*: LACFCD.
[f]*Source*: Los Angeles County Fire Department.
Note: the data appear not to be corrected for inflation to a 1974 base date, and therefore under-represent the total value of the investment.

an overall benefit : cost ratio of 1.85 : 1 (USDA 1938). Total expenditures were estimated at $22.8 million, and total benefits at $42.03 million. The cost-effectiveness of the initial programme and its subsequent developments were reviewed in 1974 (USDAFS 1974). The fire-control programmes of 1941 and 1960 were both shown to be cost-effective, yielding a benefit : cost ratio of 1.56 : 1 on an investment by 1974 of $16.7 million. The programme, it was estimated, reduced average annual burn to 82% less than might have been expected without it (59% attributable to the 1941 programme, 23% to the 1960 programme). An initial assessment of the critical area stabilisation programme (which had cost $2.5 million by 1974), and a partial analysis of the check-dam programme (a total investment by 1974 of $11 million) suggested that they were only marginally cost-effective or not cost-effective (USDAFS 1974).

Debris removal is becoming an increasingly important factor in storm costs. The approximate costs of debris removal from debris basins and reservoirs over time, at 1974 prices, is shown on Figure 4.1D. The real costs of removal are rising as debris-disposal areas become more expensive and more distant from debris sources. Partly for this reason, debris-disposal costs are increasing as a proportion of total storm-damage costs. For example, in 1969 they were approximately 20% of estimated total damage costs, but in 1978 they were 32%.

The cost-effectiveness of slope-failure-control measures is more difficult to assess. As the review of 1969 and 1978 storms showed, the effect of storms on slope failure was substantially reduced by the introduction of increasingly rigorous building controls. In 1969, for example, average damage per site for the total number of sites constructed before 1952 was $330; for sites developed between 1952 and 1962 it was $100, and for post-1963 sites it was only $7.00 (Slosson 1969). The reduction in site damage costs was only achieved at some expense, of course. For each lot, the developer incurs additional design costs ($243 on average), grading costs ($500), and city inspection costs ($335), a total of approximately $1078, or about 10% of the average losses without control (California Division of Mines and Geology 1973). It was estimated that in 1978 the benefit–cost ratio of building ordinance changes – when comparing the performance of pre-1963 and post-1963 structures – was about 25.1 : 1 (Slosson & Krohn 1979). These estimates are for private property only, but similar benefits probably accrue to public property. Leighton's (1976) analysis of the ratio of the investigative costs of preventing landsliding to the estimated losses due to landsliding, based on a study of the south-east quarter of Oat Mountain quadrangle, Los Angeles County, suggests an extremely beneficial overall ratio of 48.7 : 1 – a conclusion that points strongly towards the advantages of geological and related surveys prior to slope development.

County, state and nation

The geomorphological hazards in Los Angeles County are locally important manifestations of state and national problems. Although precise estimates are difficult to obtain for all the fluvial hazards, some comparative data are available for landslides and flooding.

The California Division of Mines and Geology (1973) estimated that the annual landslide damage in California between 1970 and 2000 would be approximately $300 million per annum, or about $9.9 billion over the whole period. Within the state, two major areas of population and urban growth in hilly terrain – the southern California metropolis, and the San Francisco Bay region – will contribute substantially to these totals. For example, in the 1969 storms the estimated total damage due to slope failure and mudflows in the former was approximately $1 billion, and in the latter it was at least $35–56 million (Sorensen *et al.* 1975, Fleming & Taylor 1980). The estimated average annual costs of such hazards per capita by 1977 were $1.60 in the City of Los Angeles, and $1.30 in the San Francisco Bay region (Fleming & Taylor 1980).

Nationally, there is little doubt that landslide damage costs hundreds of millions of dollars each year (Smith 1958). The annual cost of repair and correction of highway-related landslides alone is $100 million per annum (Chassie & Goughnour 1976). Krohn and Slosson (1976) estimated that annual landslide damage for the conterminous states on private property was about $400 million per annum; on the likely assumption that damage to public property is similar, average annual damage losses of about $1 billion are probable (Schuster 1978).

Figures of this kind are very approximate, and must be treated with caution because damage varies greatly, not only with the type of failure and local geological circumstances, but also with the type of property and the nature of the community affected. Furthermore, landslide damage is rising nationally, especially as residential growth extends into marginal land, and hilly terrain is increasingly disturbed by bulldozer, scraper and digger (Sorensen *et al.* 1975). Damage in Los Angeles County will not necessarily reflect this national trend, because the local means of control are relatively very advanced (as indeed they are in California by comparison with other states). Certainly, slope-failure damage in Los Angeles County will continue to contribute significantly to the national statistics, but its proportionate contribution may be expected to decline.

Flooding and sediment problems are also of regional and national significance. The estimated flood losses in California between 1970 and 2000 are $210 million per annum, or $6.5 billion over the whole period (California Division of Mines and Geology 1973). Southern California includes much of the potentially floodable land (State of California, Resources Agency Department of Water Resources 1978), and will probably sustain a high proportion of these losses; but the losses in Los Angeles County are not likely to increase substantially as a proportion of the regional total because of the successful implementation of flood-control programmes and the continuing extension of controls to areas of the county not originally included in these programmes. Table 4.3 shows that while storm damage losses are still absolutely and relatively high in Los Angeles County, total sustained losses as a proportion of total sustained losses *plus* estimated losses prevented by flood-control efforts are lower than in other counties in southern California.

Nationally, it has been estimated that the annual average damage caused by floods up to 1970 was about $1.2 billion (by comparison, the approximate figure for Los Angeles County is at least $4.6 million per annum between 1914 and 1970 at 1967 prices); about 10% of the population is exposed to flood threat; deaths occur at an average annual rate of two per million at risk; and the average annual costs per person at risk were about $48 (Burton *et al.* 1978). As with landslides, costs have been increasing gradually over time, from about 0.15% of gross national product in 1936 to about 0.45% in 1972. The Los

Table 4.3 Flood damage in southern California (dollar data from USACE 1969 and 1978b).

	Total losses ($ million)		Total sustained losses as a % of sustained losses plus estimated losses prevented	
County	1969	1978	1969	1978
Santa Barbara	21.9	6.7	91.0	62.6
Ventura	44.1	19.9	42.3	23.5
Los Angeles	31.4	34.5	3.0	2.7
Orange	21.9	8.7	4.7	30.3
San Diego	2.7	12.0	—	75.0
Riverside	32.1	9.0	93.3	95.2
San Bernadino	54.2	5.2	49.1	50.9

Angeles County contribution to these national statistics will probably decline proportionately in the future, because the main engineering protection schemes are approaching completion, and the county is likely to be in the vanguard of national efforts to reduce further flood losses by developing new approaches to flood-plain management through, for example, improved zoning, insurance and forecasting practices.

Sequences and cycles

Almost a century of experience has powerfully illuminated understanding of the nature of fluvial geomorphological hazards in the mountains, hills and alluvial plains of Los Angeles County. The main processes (summarised in Fig. 1.1) are primarily activated by mid-winter storms which attack a landscape that is different from that attacked by their precedessors. The effects of storms in any one season reflect the nature of the storms, the antecedent environmental changes, the contemporary spatial variations of landscape characteristics, and the diverse ways in which communities and management agencies respond to the hazards. Some of these complex circumstances are summarised in two diagrams (Figs. 4.3 & 4.5).

The first (Fig. 4.3) attempts to outline the patterns of secular change upon which a storm event is superimposed. Many of these changes are difficult to chronicle and calibrate precisely, but Figure 4.3 provides at least a general impression. Before the rapid urban expansion of the 20th century, the natural and rural landscapes of Los Angeles County were already relatively unstable in terms of geomorphological change. The geomorphological system was (and still is) characterised by features that encourage rapid change: by steep slopes and small drainage systems in the hills and mountains; complex alluvial-fan systems on the plains; highly faulted, fractured, folded and friable sediments and rapidly weathering igneous rocks; poorly protective vegetation and immature soils. Then, as now, the system was driven mainly by storms, and change was exacerbated by earthquakes and brush fires. The greatest geomorphological activity was in much the same areas as the hazardous hillside zones of today. The climatic patterns were also similar. For instance (as Fig. 4.3A & B shows), the sequence of annual precipitation and storminess (the latter exemplified by maximum daily precipitation intensity) includes highs and lows, but reveals no significant secular trends.

But several secular changes have dramatically altered the environmental conditions that successive storms encounter. Of these, population growth (Fig. 4.3C) is probably the most important index of increasing human pressure on the land. The urban expansion accompanying population growth is exemplified in Fig. 4.3 by the rise in recorded subdivision lots (D) and by the increase in property values (E). In the 20th century, land use in the county has

DOI: 10.4324/9781003168645-14

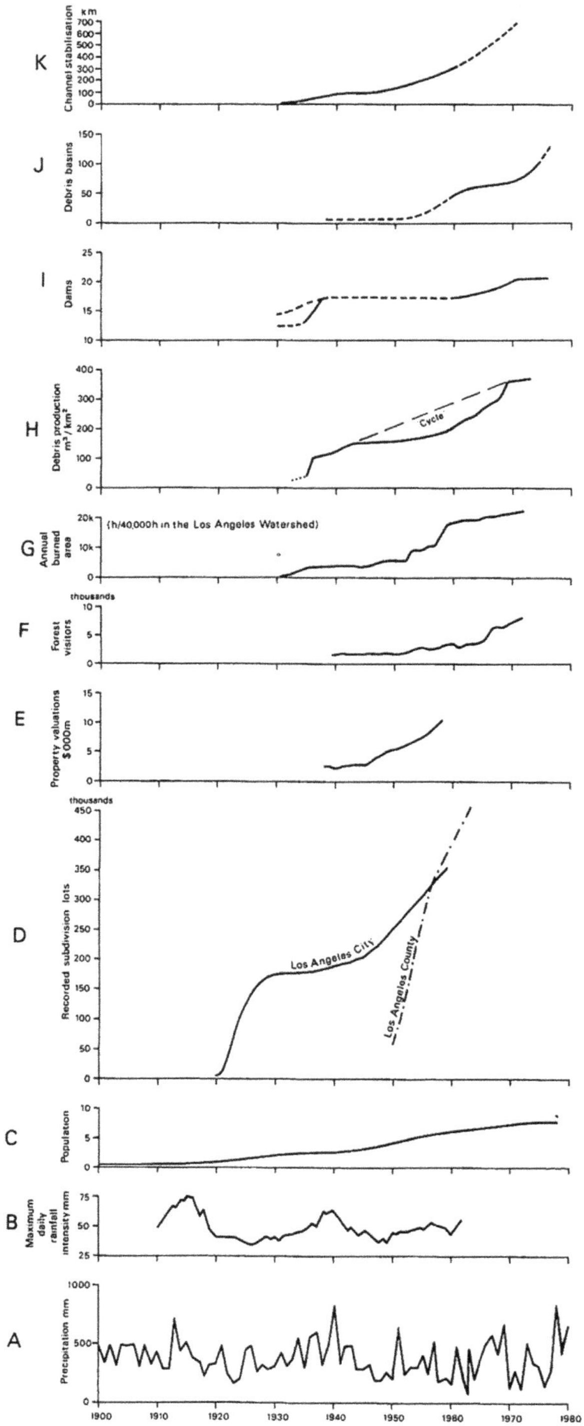
K
Channel stabilisation
km
J
Debris basins
I
Dams
H
Debris production m³/km²
Cycle
G
Annual burned area
(h/40,000h in the Los Angeles Watershed)
F
thousands
Forest visitors
E
Property valuations $000m
D
thousands
Recorded subdivision lots
Los Angeles City
Los Angeles County
C
Population
B
Maximum daily rainfall intensity mm
A
Precipitation mm
1900
1910
1920
1930
1940
1950
1960
1970
1980

seen a profound transformation as the urban frontier has spread across formerly agricultural and pastoral plains into the hills and mountains. At the same time, the growing recreational use of the mountains (Fig. 4.3F) has fuelled an increase in the annual area burned (Fig. 4.3G). This, in turn, has enhanced the influence of fire on sediment yield, although the trend of sediment yield during the period for which data are available is partly distorted by a relatively dry phase (Fig. 4.3H).

These trends serve to accelerate geomorphological change or to increase hazard potential. But such consequences have been in part and increasingly offset by the efforts of management agencies. These efforts are most tangibly expressed in the progress in building dams, debris basins and improved or stabilised channels (Figs 4.3I, J & K); as the discussion in Part 3 showed, there are many other changes – especially in management practices – that reinforce the trend towards hazard reduction and displacement. It is clearly seen in the change in flood locations as some developed places become protected, and development extends into unprotected areas (Fig. 4.4). Similarly, although more recently, management practices are alleviating the menace of urban-related slope failures, and there are changes in the locations of those failures within hillside hazard zones (see Fig. 2.9). Such trends are likely to continue, and they are likely to be accompanied by new problems and trends. For example, success in safely accommodating debris before it causes damage contributes to the reduction of flood-hazard damage, but carries with it the increasingly expensive problem of debris clearance.

Thus, each annual sequence of storms strikes a landscape and community different to some extent from those attacked by its predecessors. Yet individual storm and related post-storm events do follow a fairly predictable sequence, almost a cycle, which Figure 4.5 attempts to describe. Regional early warning of problems is provided in the months and days before the storms by

---

Figure 4.3 Selected trends in Los Angeles County. (A) Precipitation in downtown Los Angeles (data source, US Weather Bureau). (B) Maximum daily precipitation intensity based on a composite rainfall record for coastal California, ten-year moving averages (after Cooke & Reeves 1976). (C) Population growth in Los Angeles County (data source, US Department of Commerce Bureau of the Census). (D) Recorded subdivisions in Los Angeles County and the City of Los Angeles, cumulative curves (data sources, Security First National Bank 1965, Los Angeles County planning reports). (E) Total assessed property valuations, Los Angeles County (data source, USDAFS 1961). (F) Cumulative curve of visitors to the Angeles National Forest (USDAFS 1973). (G) Cumulative curve of annual burned area within the Los Angeles River Watershed in the Angeles National Forest (USDAFS 1973a). Data in hectares per 40 000 ha of forest land. (H) Cumulative debris production (m^3/km^2) in the Los Angeles River Watershed in Angeles National Forest (Ruby 1973). (I), (J) and (K) Increase in the number of dams, debris basins and kilometres of improved and stabilised channels respectively (from various sources, including Eaton 1931, LACFCD 1958, 1979a, USDAFS 1974).

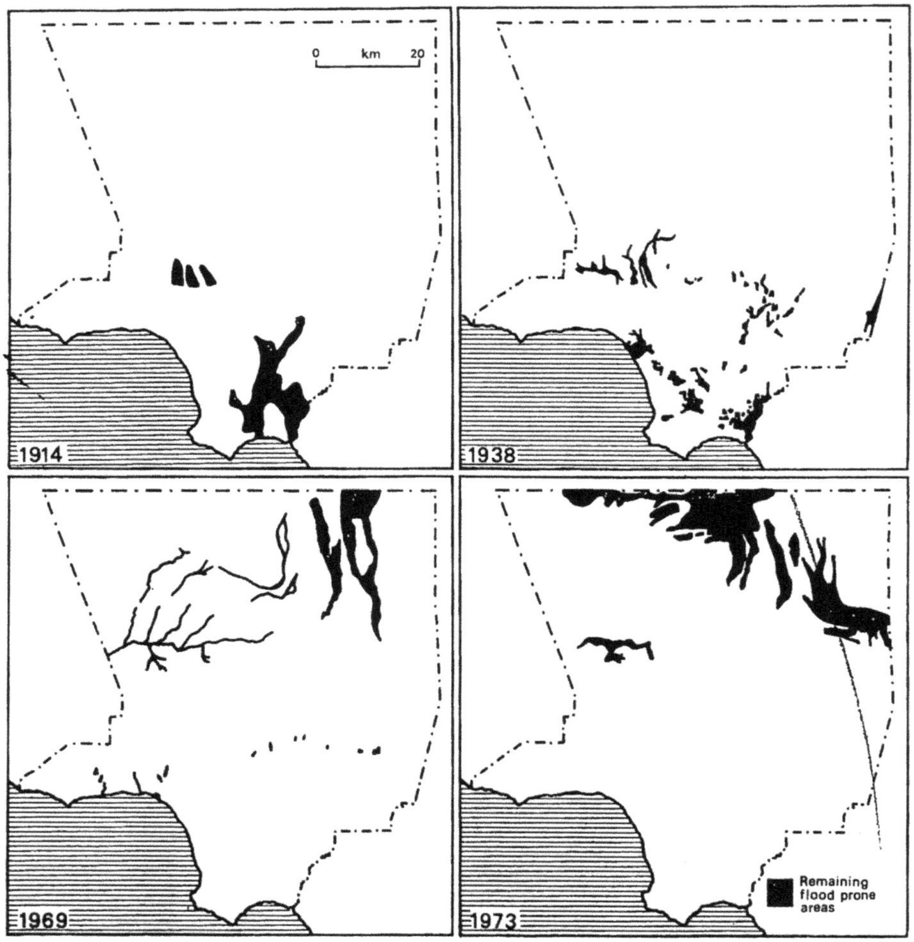

Figure 4.4 Recorded areas of flooding in Los Angeles County (1914, 1938, 1969), and the remaining flood-prone areas (1973). (After Los Angeles County Board of Supervisors 1914, LACFCD 1938, Simpson 1969, USACE 1969, Los Angeles County Regional Planning Commission 1973).

early season rainfall and weather forecasts; potentially troubled localities are intimated where land has been exposed by autumn fires and by construction activity. When the storms come, they commonly, but not always, include two phases of heavy rainfall, so that the first should act as a warning for the second. During these rainy periods, the geomorphological system may be activated to the extent that several thresholds – such as those initiating runoff, flooding, soil erosion, soil slippage and landslides – are crossed. The periods of activity above the thresholds may vary greatly from a few minutes of soil slippage to perhaps months for major landsliding. Early warning signs now normally provoke anticipatory emergency responses, such as emergency debris

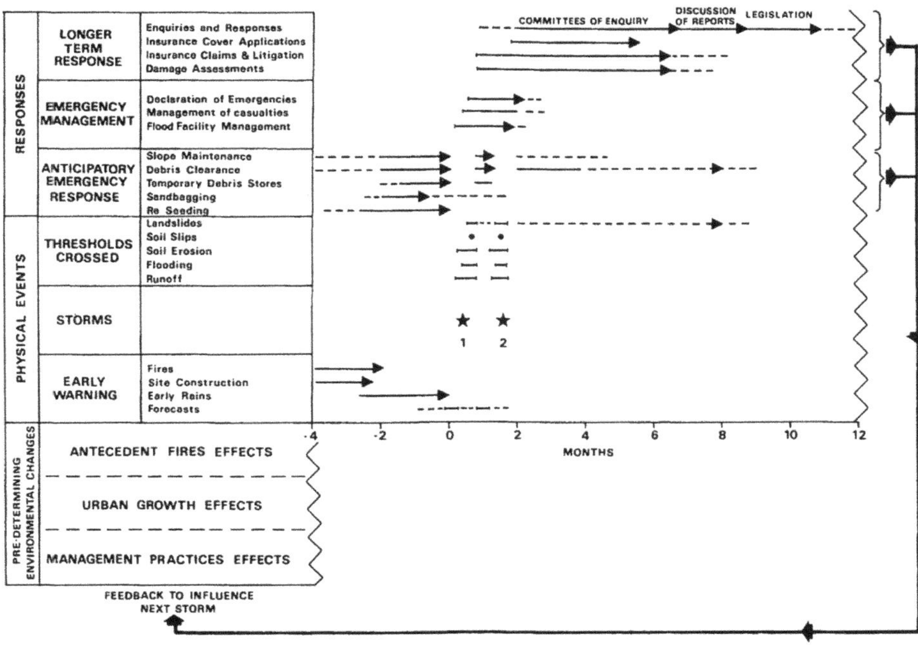

Figure 4.5 Storm hazard-response model, Los Angeles County.

clearance, the construction of temporary debris stores, sand-bagging, the re-seeding of burned areas, and slope maintenance.

Once the storm strikes, the panoply of emergency responses is initiated as required, and thresholds of response within the community that may be crossed include the declarations of local, state and federal emergencies. As the emergency subsides, the aftermath is a time of clearing up and repair, and of review, reflection and research. As insurance claims and litigation are initiated, review committees deliberate, their results are examined and, as a result, codes and practices are modified. The experience of each storm is fed back, at least in part, in an attempt to ameliorate the effects of future storms. In public agencies, experience is generally cumulative, so that over time the impact of storms of similar magnitude should decline. In general, this appears to be the case, except where urban development proceeds in advance of the effective application of control measures. For the individual, experience is also generally cumulative; but the effectiveness of individual response is reduced by the twin complications of house changes by local residents and the continuing influx of immigrants without local hazard experience.

The duration and magnitude of events in the sequence vary from one storm period to the next, and both are difficult to calibrate. For this reason, the time periods shown on Figure 4.5 are only approximate, and no attempt is made to indicate magnitude. Some differences between successive storm events mark a trend. For example, some of yesterday's Acts of God tend to become tomorrow's acts of administrative negligence, as problems are understood and responsibility for solving them is apportioned. Some of the other differences are random, perhaps reflecting a particular physical antecedent, an error, the impact of a dominant personality, or even a draconian tax cut. Nevertheless, most of the features on Figure 4.5 are common to all storms – at least those in the recent past – and they are also likely to figure in future storms.

> . . . I've seen a lot of them – storms and men. Each one is different. There are big bluffers, and the sneaks, and the honest dependable ones . . .
> . . . storms and men – they're all different, and yet they're all the same . . .
>
> (George R. Stewart 1941 *Storm*, p. 342)

References

Adelizzi, R. F. 1969. Eichler and Avner – strict liability in real property. *Los Angeles Bar Bulletin* August, 431–68.

American Institute of Professional Geologists 1966. *Report to the Honourable Sam Yorty of the committee on the geologic environment in the City of Los Angeles*. Golden: American Institute of Professional Geologists.

Anderson, H. W., G. B. Coleman and P. W. Zinke 1959. *Summer slides and winter scour . . . dry–wet erosion in southern California mountains*. USDAFS Pacific Southwest Forest and Range Experiment Station Technical Paper 36.

Association of Engineering Geologists 1966. Seminar on '*The Importance of the Earth Sciences to the Public Works and Building Official*.' Glendale: Association of Engineering Geologists.

Association of Engineering Geologists 1978. *Building Codes Committee Review – 1978 Storm Damage*. Glendale: Association of Engineering Geologists, California Section. (Includes *Storm Damage Task Force Report of the City of Los Angeles*.)

Bailey, H. P. 1954. Climate, vegetation and land use in southern California. In *Geology of southern California*, R. H. Jahns (ed.), 31–44. *California Division of Mines Bulletin* **170**.

Bailey, H. P. 1966. *The climate of southern California*, Berkeley: University of California Press.

Bailey, R. G. and R. M. Rice 1969. Soil slippage: an indicator of slope instability on chaparral watersheds in southern California. *Professional Geographer* **21**, 172–7.

Baker, R. F. and R. Chieruzzi 1959. Regional concept of landslide occurrence. *Highway Research Board Bulletin* **216**, 1–16.

Bandy, O. L. and L. Marincovich Jr 1973. Rates of Late Cenozoic uplift, Baldwin Hills, Los Angeles, California. *Science* **181**, 653–5.

Bartfield, I. and D. B. Taylor 1981. A case study of a real-time flood warning system on Sespe Creek, Ventura County, California. In *Storms, floods and debris flows in southern California and Arizona, 1978 and 1980*, 165–76. Washington DC: National Academy Press.

Bentley, J. R. 1961. *Fitting brush conversion to San Gabriel watersheds*. USDAFS Pacific Southwest Forest and Range Experiment Station Miscellaneous Paper 61.

Bigger, R. 1959. *Flood control in metropolitan Los Angeles*. Los Angeles: University of California Publications in Political Science 6.

Blanc, R. P. and G. B. Cleveland 1968. *Natural slope stability as related to geology, San Clements Area, Orange and San Diego counties, California*. California Division of Mines and Geology Special Report 98.

Bonner, W., C. Pyke and R. Hickey 1971. *A study of the California rainstorm of January 1969*. Los Angeles: Department of Meteorology, UCLA.

Borchardt, G. A. 1977. *Clay mineralogy and slope stability*. California Division of Mines and Geology Special Report 133.

Brownlie, W. R. and B. D. Taylor 1981. *Sediment management for southern California, coastal plains and shoreline. Part C, Coastal sediment delivery by major rivers in southern California*. California Institute of Technology EQL Report 17–C.

Bruington A. E. 1982. Fire-loosened sediment menaces the city. In *Proceedings of the symposium on dynamics and management of mediterranean-type ecosystems, June 22–26, 1981, San Diego, California*. USDAFS General Technical Report PSW–58, 420–2.

Brumbaugh, R. W., W. H. Renwick and L. L. Loeher 1982. Effects of vegetation change on shallow landsliding: Santa Cruz Island, California. In *Proceedings of the symposium on dynamics and management of mediterranean-type ecosystems, June 22–26, 1981, San Diego, California*. USDAFS General Technical Report PSW-58, 397–402.

Burton, I., R. W. Kates and G. F. White 1979. *The environment as hazard*. Oxford: Oxford University Press.

California Division of Mines and Geology 1969. *Geologic Map of California – Los Angeles Sheet* (1 : 250 000).

California Division of Mines and Geology 1973. Urban geology – master plan for California. *California Division of Mines and Geology Bulletin* **198**.

California Division of Mines and Geology 1979. *Landslides in the Los Angeles region, California – effects of the February–March 1978 rains*. Open-file Report.

California Geology 1971. April–May. Special San Fernando earthquake edition.

Campbell, R. H. 1973. *Isopleth map of landslide deposits, Point Dume, Los Angeles County, California*. USGS Miscellaneous Field Studies Map MF535.

Campbell, R. H. 1975. *Soil slips, debris flows and rainstorms in the Santa Monica Mountains and vicinity, southern California*. USGS Professional Paper 851.

Carpenter, F. A. 1914. Flood studies at Los Angeles. *Monthly Weather Review* **42**, 385–91.

Chassie, R. G. and R. D. Goughnour 1976. States intensifying efforts to reduce highway landslides. *Civil Engineering* **46**, 65–6.

Chawner, W. D. 1935. Alluvial fan flooding. *Geographical Review* **25**, 255–63.

City of Glendora 1969. *1969 Storm report and recommendations*.

City of Los Angeles 1967. *Landslides and attendant problems*. A report to the Mayor by the Mayor's *ad hoc* Landslide Committee, March 28, 1967.

City of Los Angeles 1968. *Ordinance 136195*, February 21, 1968.

City of Los Angeles 1972a. *Building code*. Los Angeles: Building News.

City of Los Angeles 1972b. *Planning and zoning code*. Los Angeles: Building News.

City of Los Angeles 1975. *Preliminary ordinance draft to regulate density according to the percentage of slope*. Unpublished memorandum from Director of Planning to City Planning Commission.

City of Los Angeles (no date). *Desired content of geological reports submitted to the Department of Building and Safety, City of Los Angeles*.

City of Los Angeles, Department of Building and Safety 1969. *Storm and slope failure damage report*. Submitted to Los Angeles County Flood Control District, 10 April, 1969.

City of Los Angeles, Department of Building and Safety 1978a. *Guide for erosion and debris control in hillside areas*.

City of Los Angeles, Department of Building and Safety 1978b. *Guide for prospective buyers of hillside homes*.

City of Los Angeles, Department of Building and Safety 1978c. *Proposed code changes relative to earthquake resistant design* (subsequently incorporated in part in an ordinance amending and repealing Article 1 of Chapter 9 of the Los Angeles Municipal Code).

City of Los Angeles, Department of Public Works 1969. *Administration of emergency erosion control contracts*. Special order S004–0269, February 5, 1969.

City of Los Angeles, Department of Public Works (no date). *Emergency and disaster manual*.

Cleaves, A. B. 1969. Flagg's Restaurant area landslide, Pacific Palisades, California. In G. A. Kiersche and A. B. Cleaves (eds) *Engineering geology case histories* **7**, 75–9.

Cleveland, G. B. 1971a. *Regional landslide prediction*. Sacramento: State of California, Division of Mines and Geology.
Cleveland, G. B. 1971b. The role of state governments in urban geology. In *Environmental planning and geology*, 70–5. Washington, DC: USDHUD, US Department of the Interior.
Conservation Association of Los Angeles County 1932. *Relation of watershed cover to flood control in Los Angeles County*.
Conservation Association of Los Angeles County 1938. *Plan for upstream flood control*.
Cooke, R. U. and P. Mason 1973. Desert knolls pediment and associated landforms in the Mojave Desert, California. *Revue de Géomorphologie Dynamique* **22**, 49–60.
Cooke, R. U. and R. W. Reeves 1976. *Arroyos and environmental change in the American South-west*. Oxford: Oxford University Press.
Cooke, R. U. and I. G. Simmons 1966. Some recent changes in California. *Tijdschrift voor Economische en Sociale Geografie*, November–December 1966, 232–42.
Corbett, E. S. and R. P. Crouse 1968. *Rainfall interception by annual grass and chaparral . . . losses compared*. USDAFS Pacific Southwest Forest and Range Experimental Station Research Paper PSW-48.
Corbett, E. S. and L. R. Green 1965. *Emergency revegetation to rehabilitate burned watersheds in southern California*. USDAFS Pacific Southwest Forest and Range Experiment Station Research Paper PSW-22.
Corbett, E. S. and R. M. Rice 1966. *Soil slippage increased by brush conversion*. USDAFS Forest Service Research Paper PSW-128.
Cordell, E. S. 1969. Letter from the USDA Soil Conservation Service to the Los Angeles County Flood Control District, 29th April, 1969.
Countryman, C. M. 1974. *Can southern California wildland conflagrations be stopped?* Berkeley: USDAFS Pacific Southwest Forest and Range Experiment Station.
Countryman, C. M., M. H. McCutchan and B. C. Ryan 1969. *Fire weather and the fire behavior at the 1968 canyon fire*. Berkeley: USDAFS Pacific Southwest Forest and Range Experiment Station.
Crouse, R. P. 1961. *First-year effects of land treatment of dry-season streamflow after a fire in southern California*. USDAFS Pacific Southwest Forest and Range Experiment Station Research Note 191.

Daingerfield, L. H. 1934. Excessive rain and flood in Los Angeles, California area. *Monthly Weather Review* **62**, 91–2.
Davis, J. D. 1982. Rare and unusual postfire flood events experienced in Los Angeles County during 1978 and 1980. In *Storms, floods and debris flow in southern California and Arizona, 1978 and 1980*, 243–55. Washington DC: National Academy Press.
DeBano, L. F. 1981. *Water-repellent soils: a state-of-the-art*. USDAFS Forest and Range Experiment Station General Technical Report PSW-46.
Deeming, J. E., J. W. Lancaster, M. A. Fosberg, R. W. Furman and M. J. Schroeder 1972. *National fire-danger rating system*. USDA Forest Service Research Paper RM-84.
Defense Civil Preparedness Agency 1976. *In-time of emergency – a citizen's handbook on nuclear attack, natural disasters*. Washington DC: Department of Defense.
Doehring, D. O. 1968. The effect of fire on geomorphic processes in the San Gabriel Mountains, California. *Contributions to Geology, University of Wyoming* **7**, 43–65.

Easton, W. H. 1973. Earthquakes, rain and tides at Portuguese Bend landslide, California. *Bulletin of the Association of Engineering Geologists* **10**, 173–94.
Eaton, E. C. 1931. *Comprehensive plan for flood control and conservation – present conditions and immediate needs*. Los Angeles: LACFCD.
Eaton, E. C. 1936. Flood and erosion control problems and their solution. *Transactions of the American Society of Civil Engineers* **101**, 1302–30.

Eaton, E. C. and F. Gillelen 1931. *Report on check dams*. Los Angeles: LACFCD.
Edinger, J. G., E. A. Helvey and D. Baumhefner 1964. *Surface wind patterns in the Los Angeles Basin during Santa Ana conditions. Report on Research Project 2606*. Los Angeles: Department of Meteorology, UCLA.
Ehlig, P. L. and K. W. Ehlert 1978. Engineering geology of a Pleistocene landslide in Palos Verdes. In D. L. Lamar (ed.) *Geologic guide and engineering geology case histories, Los Angeles metropolitan area*, 159–66. Los Angeles: Association of Engineering Geologists Conference 1978.
Evans, J. R. and C. H. Gray Jr (eds) 1972. Analysis of mudslide risk in southern Ventura County, California, 1971. *California Division of Mines and Geology Open-file Release* 72–23.
Evelyn, J. B. 1982. Operation and performance of Corps of Engineers flood control projects in southern California and Arizona 1978–80. In *Storms, floods and debris flows in southern California and Arizona 1978 and 1980*, 131–50. Washington DC: National Academy Press.

Ferrell, W. R. 1959. *Watershed treatment projects in the San Gabriel Mountains of Los Angeles County*. Los Angeles: LACFCD.
Fleming, R. W. and F. A. Taylor 1980. *Estimating the costs of landslide damage in the United States*. USGS Circular 832.
Foggin, G. T. and L. F. DeBano 1971. Some geographic implications of water-repellent soils. *Professional Geographer* **23**, 347–50.
Forsyth, R. A. and M. L. McCauley 1982. The Malibu slide. In *Storms, floods and debris flows in southern California and Arizona, 1978 and 1980*, 333–4. Washington DC: National Academy Press.
Fosberg, M. A. 1965. *A case study of the Santa Ana winds in the San Gabriel Mountains*. USDAFS Pacific Southwest Forest and Range Experiment Station Research Note PSW-78.
Fosberg, M. A., C. A. O'Dell and M. J. Schroeder 1966. *Some characteristics of the three-dimensional structure of Santa Ana winds*. Berkeley: USDAFS Pacific Southwest Forest and Range Experiment Station.

Garza, C. and C. Peterson 1981. Damage-producing winter storms of 1978 and 1980 in southern California: a synoptic view. In *Storms, floods and debris flows in southern California and Arizona, 1978 and 1980*, 43–56. Washington DC: National Academy Press.
Giessner, F. W. and M. Price 1971. *Flood of January 1969, near Azusa and Glendora, California*. USGS Hydrological Investigations Atlas HA424.
Gilluly, J. 1949. Distribution of mountain building in geologic time. *Bulletin of the Geological Society of America* **60**, 561–90.
Grover, N. C., H. M. McGlashan and F. F. Henshaw 1917. *Surface water supply of the United States, 1914. Part XI, Pacific slope basins of California*. USGS Water-supply Paper 391.

Hamilton, E. I. and P. B. Rowe 1949. *Rainfall interception by chaparral in California*. Berkeley: USDAFS California Forest and Range Experiment Station.
Hanes, T. L. 1971. Succession after fire in the chaparral of southern California. *Ecological Monographs* **41**, 27–52.
Hill, L. W. 1963. *The cost of converting brush cover to grass for increased water yield*. USDAFS Pacific Southwest Forest and Range Experiment Station Research Note PSW-2.
Howard, R. B. 1982. Erosion and sedimentation as part of the natural system. In *Proceedings of the symposium on dynamics and management of mediterranean-type ecosystems,*

June 22–26, 1981, San Diego, California. USDAFS General Technical Report PSW-58, 403–8.

Hughes, J. L. and A. O. Waananen 1972. *Effects of the January and February 1969 floods on ground water in central and southern California*. Menlo Park: USGS Open-file Report.

Hull, M. K., C. A. O'Dell and M. J. Schroeder 1966. *Critical fire weather patterns – their frequency and levels of fire danger*. Berkeley: USDAFS Pacific Southwest Forest and Range Experiment Station.

International Conference of Building Officials 1973. *Uniform Building Code*. Whittier: ICBO.

Jahns, R. H. (ed.) 1954. *Geology of southern California. California Division of Mines Bulletin* **170**.

Jahns, R. H. 1969. Seventeen years of response by the City of Los Angeles to geologic hazards. In *Proceedings of the Conference on Geologic Hazards and Public Problems*, R. A. Olson and M. M. Wallace (eds), 283–96. Santa Rosa: Office of Emergency Preparedness, Federal Regional Center 7.

Jahns, R. H. and K. V. Linden 1973. Space–time relationships of landsliding on the southerly side of the Palos Verdes Hills, California. In *Geology seismicity and environmental impact*, D. E. Moran, J. E. Slosson, R. O. Stone and C. A. Yelverton (eds), 123–38. Los Angeles: Association of Engineering Geologists Special Publication.

Kasperson, R. E. 1969. Environmental stress and the municipal political system: the Brockton water crisis of 1961–1966. In *The structure of political geography*, R. E. Kasperson and J. V. Minghi (eds), 481–96. London: University of London Press.

Kenyon, E. C. Jr and R. J. Coakley 1960. Sedimentation data on flood control reservoirs, debris dams and debris basins. Los Angeles: LACFCD.

Kenyon, E. C. and W. Jones 1959. Appendix to *Fire frequency study*. Los Angeles: LACFCD.

Kirkby, M. J. and R. P. C. Morgan (eds) 1980. *Soil erosion*. Chichester: Wiley.

Knott, J. M. 1980. *Reconnaissance assessment of erosion and sedimentation in the Cañada de los Alamos Basin, Los Angeles and Ventura counties, California*. USGS Water-supply Paper 2061.

Kockelman, W. J. 1980. *Tools to avoid landslide hazards and reduce damage*. USGS Open-file Report 80–487.

Kraebel, C. J. 1934. The La Crescenta flood. *American Forests* **40**, 251–4.

Krammes, J. S. 1960. *Erosion from mountainside slopes after fire in southern California*. USDAFS Research Note 171.

Krammes, J. S. 1963. *Seasonal debris movement from steep mountainside slopes in southern California*. In *Proceedings of the Federal Inter-Agency Sedimentation Conference 1963*. USDA Miscellaneous Publication 970, 85–8.

Krammes, J. S. and L. F. DeBano 1965. Soil wettability: a neglected factor in watershed management. *Water Resources Research* **1**, 283–6.

Krammes, J. S. and H. Hellmers 1963. Tests of chemical treatments to reduce erosion from burned watersheds. *Journal of Geophysical Research* **68**, 3667–72.

Krammes, J. S. and J. Osborn 1969. Water-repellent soils and wetting agents as factors influencing erosion. In *Proceedings of the symposium on water-repellent soils*, 177–87. Riverside: University of California.

Krohn, I. P. and J. E. Slosson 1976. Landslide potential in the U.S. *California Geology* **29**, 224–31.

Kunreuther, H. 1968. The case for comprehensive disaster insurance. *Journal of Law and Economics* **11**, 133–63.

Langbein, W. B. 1953. Flood insurance. *Journal of Land Economics* **29**, 323–30.
Larson, J. H. 1966. Legal aspects of earth sciences. In *Seminar on 'The Importance of the Earth Sciences to the Public Works and Building Official'*, 409–48. Glendale: Association of Engineering Geologists.
Leighton, F. B. 1966. Landslides and hillslope development. In *Engineering geology in southern California*, Lung, R. and R. Proctor (eds), 147–204. Arcadia: Association of Engineering Geologists.
Leighton, F. B. 1969. Landslides. In *Proceedings of the Conference on Geologic Hazards and Public Problems*, R. A. Olson and M. M. Wallace (eds), 97–132. Santa Rosa: Office of Emergency Preparedness Federal Regional Centre 7.
Leighton, F. B. 1971. The role of the consulting geologist in urban geology. In *Environmental planning and geology*, 82–9. Washington DC: USDHUD, and US Department of the Interior.
Leighton, F. B. 1972. Origin and control of landslides in the urban environment of California. *24th International Geological Congress, Section 13*, 89–96.
Leighton, F. B. 1976. *Urban landslides: targets for land-use planning in California*. Geological Society of America Special Paper 174, 37–60.
Leopold, L. B. and T. Maddock Jr 1954. *The flood control controversy*. New York: Ronald Press.
Li, R. M. *et al.* 1979. *Hydraulic model study of flow structures: Contract No. 16-712-CA/4*. Fort Collins: Department Civil Engineering, Colorado State University.
Lipkis, A., S. Hough and L. N. Geller 1981. Flood assistance on private property: private response to a government responsibility? In *Storms, floods and debris flows in southern California and Arizona, 1978 and 1980*, 477–82. Washington, DC: National Academy Press.
Los Angeles Chamber of Commerce, Joint Technical Committee on Earthquake Protection 1933. *Earthquake hazard and earthquake protection*.
Los Angeles County 1968. *Building laws*. Los Angeles: Building News.
Los Angeles County 1970. *Los Angeles County and Cities disaster relief manual*. Los Angeles County and Cities Disaster and Civil Defense Commission.
Los Angeles County (no date). *Minimum standards for engineering geology reports submitted to the Engineering Geology section for review*.
Los Angeles County Board of Supervisors 1914. *Provisional Report of the Board of Engineers Flood Control, June 3*.
Los Angeles County Board of Supervisors 1915. *Report of the Board of Engineers Flood Control, July 27*.
Los Angeles County Board of Supervisors 1971. *Report of the Los Angeles County Earthquake Commission, San Fernando Valley earthquake, February 9, 1971*.
Los Angeles County Board of Supervisors 1972. *Report by Earthquake Task Force "A" concerning the recommendations of the Los Angeles County Earthquake Commission*.
Los Angeles County Board of Supervisors 1974. *General plan program – environmental development guide*.
Los Angeles County Department of County Engineer 1973. *Erosion potential from the June 22, 1973, fire*.
Los Angeles County Department of Forester and Fire Warden 1938. *Report covering flood damage of February–March 1938 to watershed fire protection facilities in the County of Los Angeles, California to Los Angeles County Board of Supervisors*.
Los Angeles County Department of Regional Planning 1974. *Los Angeles County general plan, seismic safety element*. Proposed element, draft environmental impact report, and other reports.
Los Angeles County Earthquake Commission 1971. *San Fernando earthquake, February 9, 1971*.
Los Angeles County Engineer 1965. *Current practices relative to control of development in*

areas subject to geologic hazards. Unpublished memorandum to the County Board of Supervisors.

Los Angeles County Engineer 1970. *County Engineer policy for geologic review and approval of tentative subdivision maps and parcel maps*.

Los Angeles County Engineer and the Los Angeles County Flood Control District 1970. *Antelope Valley. Flood control and water conservation . . . a plan of improvement*.

Los Angeles County Fire Department 1967a. *Emergency operations procedure*.

Los Angeles County Fire Department 1967b. *County of Los Angeles fire code*. County Ordinance 2947.

Los Angeles County Fire Department 1969. *Flood report, January 19–26, 1969*.

Los Angeles County Flood Control District 1931. *Possible overflow areas during flood 50% greater than 1914 under present conditions* (map).

Los Angeles County Flood Control District 1934. *Report of S. M. Fisher, Chief Engineer, on control of flood, storm and other waste waters of the district*.

Los Angeles County Flood Control District 1935. *Engineer's estimate of cost, La Crescenta and Montrose area*.

Los Angeles County Flood Control District 1938. *Flood of March 2, 1938*.

Los Angeles County Flood Control District 1952. *Report of flood of January 15–18, 1952*.

Los Angeles County Flood Control District 1958. *Report on channel stabilization feasibility*.

Los Angeles County Flood Control District 1959. *Report on debris reduction studies*. (The 'blue book'.)

Los Angeles County Flood Control District 1961. *Waste water reclamation in Los Angeles County*.

Los Angeles County Flood Control District 1963. *Isohyetal map showing 90-year normal (1872–1962) seasonal precipitation for Los Angeles County*.

Los Angeles County Flood Control District 1964. *Storm operation guide*.

Los Angeles County Flood Control District 1965. *Debris dams and basins. Criteria for design*.

Los Angeles County Flood Control District 1969. *Storm damage report: Sant Anita and Sawpit canyons channel stabilization project*.

Los Angeles County Flood Control District 1970a. *Report on the flood hazards resulting from fires occurring within the Los Angeles County drainage boundaries, September 25–October 2, 1970*.

Los Angeles County Flood Control District 1970b. *Report on the control of surface storm water by storm drains and drainage channels*.

Los Angeles County Flood Control District 1976. *Maintenance management system study, design report, emergency response element*.

Los Angeles County Flood Control District 1979a. *Biennial report 1977–79*.

Los Angeles County Flood Control District 1979b. *Engineering methodology for mudflow analysis*.

Los Angeles County Flood Control District 1981. *Procedure for determining flood protection areas and flood proofing elevations*.

Los Angeles County Flood Control District (no date). *Subdivision manual*.

Los Angeles County Regional Planning Commission 1972. *Los Angeles County zoning ordinance*.

Los Angeles County Regional Planning Commission 1973. *General plan of Los Angeles County*.

Lustig, L. K. 1965. *Sediment yield of the Castaic watershed, western Los Angeles County, California – a quantitative geomorphic approach*. USGS Professional Paper 422-F.

Lynch, H. B. 1931. *Rainfall and stream run-off in southern California since 1769*. Los Angeles: Metropolitan Water District of southern California.

Lynch, H. B. 1948. Pacific Coast rainfall – wide fluctuations in a hundred years! *Western Construction News*, 76–80.

McGill, J. T. 1959. *Preliminary map of landslides in the Pacific Palisades area, City of Los Angeles, California*. USGS Miscellaneous Geologic Investigations Map I-284.

McGill, J. T. 1973. *Map showing landslides in the Pacific Palisades area, City of Los Angeles, California*. USGS Miscellaneous Field Studies Map MF-471.

Martinez, A. 1978. Winter's devastating storms – the impact lingers on. *Los Angeles Times*, pt II, Sunday, July 2, 1978.

Maxwell, J. C. 1960. *Quantitative geomorphology of some mountain chaparral watersheds of southern California*. Columbia University Department of Geology Technical Report 19.

Merriam, R. 1960. Portuguese Bend landslide, Palos Verdes Hills, California, *Journal of Geology* **68**, 140–53.

Miller, W. J. 1957. *California through the ages*. Los Angeles: Westernlore Press.

Mitchell, B. 1979. *Geography and resource analysis*. London: Longman.

Morgan, R. P. C. (ed.) 1980. *Soil conservation: problems and prospects*. Chichester: Wiley.

Morton, D. M. 1971. Seismically triggered landslides in the area above the San Fernando Valley. In *The San Fernando, California, earthquake of February 9, 1971*. USGS Professional Paper 733, 99–103.

Morton, D. M. and R. Streitz, 1967. Landslides. *California Geology* **20**, 123–9, 135–40.

Morton, D. M. and R. Streitz 1969. *Preliminary reconnaissance map of major landslides, San Gabriel Mountains, California*. California Division of Mines and Geology Map Sheet 15.

National Academy of Sciences 1969. *Slope protection for residential developments*. Washington DC: Building Research Advisory Board Study for Federal Housing Administration, National Academy of Sciences.

National Research Council. Committee on Natural Disasters, Commission on Sociotechnical Systems 1982. *Storms, floods and debris flows in southern California and Arizona, 1978 and 1980*. Proceedings of a Symposium, September 17–18, 1980. Washington DC: National Academy Press.

Nelson, H. J. 1959. The spread of an artificial landscape over southern California. *Annals of the Association of American Geographers* **49**, Supplement, 80–100.

Nelson, H. J. and W. A. V. Clark 1976. *Los Angeles – the metropolitan experience*. Cambridge, Mass: Ballinger.

Oakeshott, G. B. 1973. 40 years ago – the Long Beach–Compton earthquake of March 10, 1933. *California Geology* **26**, 55–9.

Office of Emergency Preparedness 1969. *Proceedings of the Conference on Geological Hazards and Public Problems*, R. A. Olson and M. M. Wallace (eds). Santa Rosa: Office of Emergency Preparedness, Federal Regional Centre 7.

Orme, A. R. and R. G. Bailey 1970. The effect of vegetation conversion and flood discharge on stream channel geometry: the case of southern California watersheds. *Association of American Geographers Proceedings* **2**, 101–6.

Orme, A. R. and R. G. Bailey 1971. Vegetation conversion and channel geometry in Monroe Canyon, southern California. *Association of Pacific Coast Geographers Yearbook* **33**, 65–82.

Palmer, L. 1976. Application of land-use constraints on Oregon. In *Urban geomorphology*, D. R. Coates (ed.). Geological Society of America, Special Paper 174, 61–84.

Patton, J. H. Jr 1973. The engineering geologist and professional liability. In *Geology, seismicity, and environmental impact*, D. E. Moran, J. E. Slosson, R. O. Stone and C. A. Yelverton (eds), 5–8. Los Angeles: Association of Engineering Geologists, Special Publication.

Patric, J. H. 1959. Increasing water yield in southern California mountains. *Journal of the American Water Works Association* **51**, 474–80.

Pestrong, R. 1976. Landslides – the descent of Man. *California Geology* **29**, 147–51.

Phillips, C. B. 1971. *California aflame!* Sacremento: State of California, The Resources Agency, Department of Conservation, Division of Forestry.

Pipkin, B. W. and M. Ploessel (no date). *Coastal landslides in southern California*. Los Angeles: Department of Geological Sciences, University of Southern California, Sea Grant Publication.

Preston, R. E. 1967. Urban development in southern California between 1940 and 1965. *Tijdschrift voor Economische en Sociale Geografie*, Sept.–Oct. 1967, 237–54.

Putnam, W. C. 1942. Geomorphology of the Ventura region, California. *Bulletin of the Geological Society of America* **53**, 691–754.

Radbruch, D. H. and K. C. Crowther 1973. *Map showing areas of estimated relative amounts of landslides in California*. USGS Miscellaneous Geologic Investigations Map I-747.

Rantz, S. E. 1970. *Urban sprawl and flooding in southern California*. USGS Circular 601-B.

Reagan, J. W. 1915, February 27. Letter to Chairman of the Los Angeles County Board of Engineers on public opinion relative to mountain reservoirs.

Reagan, J. W. 1917. *Report of the Los Angeles County Flood Control District*. Filed with the Board of Supervisors of Los Angeles County, and adopted, January 2, 1917.

Retzer, J., J. E. Davis, R. S. Dalen, M. F. Doyle III, I. Sherman, R. Spencer and R. W. Trygar 1951. *The origin and movement of sediments in the Los Angeles River watershed, California*. USDAFS California Forest and Range Experiment Station. (Second draft typing.)

Rice, R. M. 1973. The hydrology of chaparral watershed. *Proceedings of the Symposium on Living with the Chaparral*, held at the Sierra Club, Riverside, California, March 30–31, 1973, 27–34.

Rice, R. M., R. P. Crouse and E. S. Corbett 1963. Emergency measures to control erosion after a fire on the San Dimas Experimental Forest. In *Proceedings of the Federal Inter-Agency Sedimentation Conference, 1963*. USDA, Miscellaneous Publication 970, 886–97.

Rice, R. M., R. R. Ziemer and S. C. Hankin 1982. Slope stability effects of fuel management strategies – inferences from Monte Carlo simulation. In *Proceedings of the symposium on dynamics and management of mediterranean-type ecosystems, June 22–26, 1981, San Diego, California*. USDAFS, General Technical Report PSW-58, 365–71.

Richards, C. A. 1970. How the geologist can help your city. *The American City*, June 1970, 84–6.

Richards, K. 1982. *Rivers*. London: Methuen.

Richter, C. F. 1954. Earthquakes and earthquake damage in southern California. In *Geology of southern California*, R. H. Jahns (ed.). *California Division of Mines and Geology Bulletin* **170**, 5–10.

Rogers, M. J. 1982. Fire management in southern California. In *Proceedings of the symposium on dynamics and management of mediterranean-type ecosystems, June 22–26 1981, San Diego, California*. USDAFS General Technical Report PSW-58, 496–501.

Root, A. W. 1958. Prevention of landslides. In *Landslides and engineering practice*, E. B. Eckel (ed.). Highway Research Board Special Report 29, 113–49.

Rowe, P. B., C. M. Countryman and H. C. Storey 1954. *Hydrologic analysis used to determine effects of fire on peak discharge and erosion rates in southern California watersheds*. Berkeley: USDAFS California Forest and Range Experiment Station.

Rowe, P. B. and L. F. Reimann 1961. Water use by brush, grass and grass-forb vegetation. *Journal of Forestry* **59**, 175–81.

Ruby, E. C. 1973. *Sediment trend study*. Pasadena: USDAFS.

Ruby, E. C. 1976. Evaluation of an extensive sediment control effort in the Los Angeles River Basin. *Proceedings of the Third Federal Inter-Agency Sedimentation Conference*, 2.91–2.102.

Russell, H. E. 1969. Letter from the American National Red Cross (Los Angeles chapter) to the Los Angeles County Flood Control District, 14 February 1969.

Ryan, B. C., G. R. Ellis and D. V. Lust 1971. *Low-level wind maxima in the 1969 San Mateo and Walker fires*. Berkeley: USDAFS Pacific Southwest Forest and Range Experiment Station.

Schroeder, M. J. 1964. *Synoptic weather types associated with critical fire weather*. Berkeley: USDAFS Pacific Southwest Forest and Range Experiment Station.

Schulman, E. 1956. *Dendroclimatic changes in semi-arid America*. Tucson: University of Arizona Special Publication.

Schumm, S. A. 1963. *The disparity between present rates of denudation and orogeny*. USGS Professional Paper 454-H.

Schuster, R. L. 1978. Introduction. In *Landslides – analysis and control*, Schuster, R. L. and R. J. Krizet (eds) National Academy of Sciences, Transport Research Board Special Report 176, 1–10.

Scott, K. M. 1971. *Origin and sedimentology of 1969 debris flows near Glendora, California*. USGS Professional Paper 750-C, C242–C247.

Scott, K. M. 1973. *Scour and fill in Tujunga Wash – a fanhead valley in urban southern California – 1969*. USGS Professional Paper 732-B.

Scott, K. M. and R. P. Williams 1978. *Erosion and sediment yields in the Transverse Ranges, southern California*. USGS Professional Paper 1030.

Scullin, C. M. 1966. History, development and administration of excavation and grading codes. In *Engineering geology in southern California*, Lung, R. and R. Procter (eds), 227–36. Arcadia: Association of Engineering Geologists.

Security First National Bank 1965. *Southern California report*.

Sharp, R. P. 1954. Some physiographic aspects of southern California. In *Geology of southern California*, R. H. Jahns (ed.), 5–10. *California Division of Mines Bulletin* **170**.

Sharp, R. P. and L. H. Nobles 1953. Mudflow of 1941 at Wrightwood, southern California. *Geological Society of America Bulletin* **64**, 547–60.

Simpson, L. D. 1969. *Hydrologic report on the storms of 1969*. Los Angeles: LACFCD.

Sinclair, J. D. 1954. Erosion in the San Gabriel Mountains of California. *Transactions of the American Geophysical Union* **35**, 264–8.

Skafte, J. E. 1930. *A theoretical study of the value of small dams as flood control structures*. Los Angeles: LACFCD.

Slosson, J. E. 1969. *The role of engineering geology in urban planning. The Governor's conference on environmental geology*. Colorado Geological Survey Special Publication 1, 8–15.

Slosson, J. E. and J. P. Krohn 1979. AEG building code review: mudflows/debris flow damage February 1978 storm, Los Angeles area. *California Geology* **32**, 8–11.

Slosson, J. E. and J. P. Krohn 1982. Southern California landslides of 1978 and 1980. In *Storms, floods and debris flows in southern California and Arizona, 1978 and 1980, 291–304*. National Research Council. Washington DC: National Academy Press.

Smalley, J. P. 1971. *Geologic parameters of slope stability in the Upper Santa Yñez drainage system*. USDAFS unpublished report.

Smalley, J. P. and J. Cappa 1971. *Reconnaissance of geologic hazards related to the Romero Burn*. USDAFS Los Padres National Forest.

Smith, K. and G. A. Tobin 1979. *Human adjustment to the flood hazard*. London: Longman.

Smith, R. 1958. Economic and legal aspects. In *Landslides and engineering practice*, E. B. Eckel (ed.). National Academy of Sciences, Highway Research Board Special Report 29, 6–19.

Smith, T. W. 1966. Repair of landslides. In *Association of Engineering Geologists Seminar on 'The Importance of the Earth Sciences to the Public Works and Building Official'*, 205–45. Glendale: Association of Engineering Geologists.

Sorensen, J. H. and N. J. Eriksen and D. S. Mileti 1975. *Landslide hazards in the United States – a research assessment*. Boulder: Institute of Behavioural Science, University of Colorado.

State of California 1971a. *Senate Bill 682*. (An act relating to uncompleted state highway facilities damaged by earthquake.)

State of California 1971b. *Assembly Bill 2210*. (An act to amend earthquake protection bills.)

State of California Department of Conservation 1972. *Task force on California's wildland fire problem: recommendations to solve California's wildland fire problem*. Sacramento: State of California, Resources Agency, Department of Conservation.

State of California Office of Emergency Services 1974. *Guidance for the development of a local emergency plan*.

State of California Office of the Governor 1977a. *Proceedings of the Governor's drought conference, March 7–8, 1977*.

State of California Office of the Governor 1977b. *The California drought 1977, an update*, February 15, 1977.

State of California Office of the Governor 1978. Memorandum from Thom. Dederer (Governor's Office, Office of Planning and Research) on mudslides, dated March 21, 1978.

State of California The Resources Agency 1965. *Landslides and subsidence – geologic hazards conference*. State of California, Los Angeles, 26–27 May, 1965.

State of California The Resources Agency, Department of Conservation 1971. *Environmental impact of urbanization on the foothill and mountainous lands of California*.

State of California Resources Agency, Department of Water Resources 1978. *California flood management: an evaluation of flood damage and prevention programs. Review Draft*. Bulletin, 199.

Stewart, G. R. 1941. *Storm*. New York: Random House.

Stone, R. 1961. *Geologic and engineering significance of changes in elevation revealed by precise levelling, Los Angeles area, California* (abs.). Geological Society of America Special Paper 68, 57–8.

Stout, M. L. 1969. *Radiocarbon dating of landslides in southern California and engineering geology implications*. Geological Society of America Special Paper 123, 167–79.

Strahler, A. N. 1950. Equilibrium theory of erosional slopes approached by frequency distribution analysis. *American Journal of Science* **248**, 673–96; 800–14.

Tatum, F. E. 1963. A new method of estimating debris-storage requirements for debris basins. In *Proceedings of the Federal Inter-agency Sedimentation Conference 1963*. USDA Miscellaneous Publication 970, 886–97.

Taylor, C. A. 1934. Debris flow from canyons in Los Angeles County Flood. *Engineering News-Record* April 5, 439–40.

Troxell, H. C. and J. Q. Peterson 1937. *Flood in La Cañada Valley, California, January 1, 1934*. USGS Water-supply Paper 796-C.

Troxell, H. C. *et al.* 1942. *Floods of March 1938 in southern California*. USGS Water-supply Paper 844.

Turhollow, A. F. 1975. *A history of the Los Angeles District, U.S. Army Corps of Engineers.* Los Angeles: US Army Engineers District.

United States Army Corps of Engineers 1938. *Flood control in the Los Angeles County drainage area.* Los Angeles: Engineer Office.

United States Army Corps of Engineers 1943. *Benefits from flood control.* Los Angeles: US Army Corps of Engineers, Los Angeles District.

United States Army Corps of Engineers 1969. *Report on floods of January and February 1969 in southern California.* Los Angeles District: US Army Corps of Engineers. (Includes Appendix D, on Los Angeles County.)

United States Army Corps of Engineers 1976. *Pacific Palisades area, Los Angeles County, California – report on landslide study*, 2 vols. Los Angeles District: US Army Corps of Engineers.

United States Army Corps of Engineers 1978a. *Natural disaster activities.* Los Angeles: US Army District.

United States Army Corps of Engineers 1978b. *Report on floods of February and March 1978 in southern California.* Los Angeles District: US Army Corps of Engineers.

United States Army Corps of Engineers 1982. *The Los Angeles County drainage area (review) study (LACDA), Fact Sheet.* Los Angeles: USACE.

United States Department of Agriculture 1938. *Survey report: Los Angeles River Watershed.* (Program for run-off and waterflow retardation and soil erosion prevention in aid of flood control.) Field Flood Control Coordinating Committee, no. 18.

United States Department of Agriculture 1940. *Review of survey report . . . by the Department of Agriculture Field Flood Control Coordinating Committee, No. 18.* The State of California Division of Water Resources and the College of Agriculture, University of California.

United States Department of Agriculture 1941. *Memorandum of understanding between the Secretary of Agriculture of the United States and the County of Los Angeles (and others).* Also House of Representatives 1941, 77th Congress, *Document 426.*

United States Department of Agriculture 1953. *Aspects of erosion and sedimentation in the 1952 floods, San Fernando Valley, Los Angeles County, California.* USDA Soil Conservation Service, Pacific Region.

United States Department of Agriculture 1957. *The Malibu fires, Los Angeles and Ventura counties, December 26 to 30, 1956.* Los Angeles: California Forest and Range Management Station.

United States Department of Agriculture Forest Service 1939. *Survey reports – Los Angeles River: partial plan for run-off and waterflow retardation and soil erosion prevention for flood control purposes (Arroyo Seco flood source area).* Field Flood Control Coordinating Committee No. 18.

United States Department of Agriculture Forest Service 1947. *Special report, channel improvement program, January 20, 1947.*

United States Department of Agriculture Forest Service 1953. *Santa Clara–Ventura Rivers and Calleguas Creek watershed: report of survey.*

United States Department of Agriculture Forest Service 1954. Fire–flood sequences on the San Dimas Experiment Forest. *USDAFS California Forest and Range Experiment Station Technical Paper* **6**.

United States Department of Agriculture Forest Service 1961. *Los Angeles River flood prevention project, supplement to review report.*

United States Department of Agriculture Forest Service 1969. *Report on storm damage caused by the January 18–26, 1969 storms and needs to protect and restore watershed facilities on the Angeles National Forest and the state and private wildlands of Los Angeles County.*

United States Department of Agriculture Forest Service 1973a. *Analysis of fire protection effectiveness 1941–1971*. Pasadena: USDAFS, Los Angeles River Watershed, Angeles National Forest.

United States Department of Agriculture Forest Service 1973b. *Evaluation of check dams for sediment control*. Pasadena: USDAFS, Angeles National Forest.

United States Department of Agriculture Forest Service 1974. *Los Angeles River flood prevention project, mountain and foothill area, review report*.

United States Department of Agriculture Forest Service 1979a. *Review report and environmental assessment Los Angeles River flood prevention project*.

United States Department of Agriculture Forest Service 1979b. *Los Angeles River flood prevention project review report*.

United States Department of Agriculture Forest Service 1979c. *A watershed management report, Angeles National Forest, 1977* (7 volumes). Los Angeles: USDAFS.

United States Department of Agriculture Forest Service 1982. *Review report – Los Angeles River flood prevention project*.

United States Department of Commerce 1975. Bureau of the Census. *Historical statistics of the United States (House of Representatives, Document 93–78)*.

United States Department of Commerce 1977. *Guide for flood and flash flood preparedness planning*. Silver Spring: National Oceanic and Atmospheric Administration.

United States Department of Housing and Urban Development 1974. *National flood insurance program*. Washington DC: USDHUD.

United States Department of Housing and Urban Development 1977a. *How to read a flood insurance rate map*. Washington, DC: USDHUD.

United States Department of Housing and Urban Development 1977b. *Flood insurance study – guidelines and specifications for study contractors*. Washington, DC: USDHUD Federal Insurance Administration.

United States Department of Housing and Development 1977c. *Flood insurance study – City of San Dimas, California*. Washington, DC: USDHUD Federal Insurance Administration.

United States Department of Housing and Urban Development/US Department of the Interior 1971. *Environmental planning and geology*. Washington, DC: USDHUD.

United States House of Representatives 1940. *Los Angeles and San Gabriel rivers and their tributaries, and Ballona Creek, California*. 76th Congress, *Document 838*. Washington, DC: US Government Printing Office.

United States Weather Bureau 1966. *Climates of the States – California*, 60–4. Washington, DC: US Government Printing Office.

Van Arsdol, M. D. 1969. Population and the United States urban environment, the Los Angeles case. *International Union for the Scientific Study of Population Conference Proceedings* A10.21–A10.212.

Van Arsdol, M. D., G. Sabagh and F. Alexander 1964. Reality and perception of environmental hazards. *Journal of Health and Human Behaviour* **5**, 144–53.

Van Arsdol, M. D., G. Sabagh and F. Alexander 1975. *Hazards of the metropolitan environment, the Los Angeles case*. Unpublished manuscript.

Varnes, D. 1958. Landslide types and processes. In *Landslides in engineering practice* NAS–NRC Highways Research Report 29, E. B. Eckel (ed.), 20–47. Washington, DC: National Academy of Sciences.

Waananen, A. O. 1969. *Floods of January and February 1969 in central and southern California*. Menlo Park: USGS Open-file Report.

Waldrep, J. S. 1966. The relationship of building codes to the earth sciences. In seminar on '*The Importance of the Earth Sciences to Public Works and Building Official*', 387–408. Anaheim: Association of Engineering Geologists.

Ward, R. 1978. *Floods*. London: Macmillan.

Weber, F. H. Jr 1982. Landsliding and flooding in southern California during the winter of 1979–80. In *Storms, floods and debris flows in southern California and Arizona, 1978 and 1980, 321–31*. Washington, DC: National Academy Press.
Weide, D. L. 1968. *The geography of fire in the Santa Monica Mountains*. MS thesis, California State College, Los Angeles.
White, G. F. and J. E. Haas 1975. *Assessment of research on natural hazards*. Mass.: M.I.T. Press.
Wright, R. H., R. H. Campbell and T. H. Nilsen 1974. Preparation and use of isopleth maps of landslide deposits. *Geology* **2**, 483–5.

Yelverton, C. A. 1971. The role of local government in urban geology. In *Environmental planning and geology*, 76–81. Washington, DC: USDHUD, US Department of the Interior.
Yelverton, C. A. 1972. *Processing tentative tract maps, private street and parcel maps*. Los Angeles: City of Los Angeles, Department of Building and Safety.
Yelverton, C. A. 1973. Land failure insurance. In *Geology, seismicity and environmental impact*, D. E. Moran, J. E. Slosson, R. O. Stone and C. A. Yelverton (eds), 15–20. Los Angeles: Association of Engineering Geologists, Special Publication.

Federal Public Laws cited in the text

PL 74-738	June 22, 1936, c. 688, sec. 1–9, 49 Stat. 1570; U.S.C. Title 33, sec. 701a–f.
PL 77-377	Dec. 26, 1941, c. 629, 55 Stat. 862; U.S.C. Title 31, sec. 80b.
PL 81-875	Sept. 30, 1950, c. 1125, sec. 1–8, 64 Stat. 1109; U.S.C. Title 42, sec. 1855a–g.
PL 81-920	Jan. 12, 1951, c. 1228, sec. 1, 64 Stat. 1245; U.S.C. Title 50 App., sec. 2251.
PL 83-566	Aug. 4, 1954, c. 656, sec. 1–12, 68 Stat. 666; U.S.C. Title 16, sec. 1001–8.
PL 84-99	June 28, 1955, c. 194, 69 Stat. 186; U.S.C. Title 33, sec. 701n.
PL 90-488	Aug. 15, 1968, 82 Stat. 770; U.S.C. Title 7, sec. 1923.
PL 91-606	Dec. 31, 1970, sec. 1, 84 Stat. 1744; U.S.C. Title 42, sec. 4401.
PL 93-234	Dec. 31, 1973, sec. 1, 87 Stat. 975; U.S.C. Title 42, sec. 4002.
PL 93-288	May 22, 1974, sec. 1, 88 Stat. 143; U.S.C. Title 42, sec. 5121.

Index

Numbers in *italics* refer to text figures. Numbers in **bold** type refer to tables.

For Product Safety Concerns and Information please contact our EU representative GPSR@taylorandfrancis.com
Taylor & Francis Verlag GmbH, Kaufingerstraße 24, 80331 München, Germany

www.ingramcontent.com/pod-product-compliance
Ingram Content Group UK Ltd.
Pitfield, Milton Keynes, MK11 3LW, UK
UKHW020201200926
475702UK00004B/564
9780367768126